T0281298

CHEMICAL GRAPH THEORY

Mathematical Chemistry

A series of books edited by:
Danail Bonchev, Department of Physical Chemistry, Higher Institute of Chemical
   Technology, Burgas, Bulgaria
Dennis H. Rouvray, Department of Chemistry, University of Georgia, Athens,
   Georgia, USA

Volume 1
CHEMICAL GRAPH THEORY: Introduction and Fundamentals
Edited by Danail Bonchev and Dennis H. Rouvray

Volumes in preparation

Volume 2
CHEMICAL GRAPH THEORY: Reactivity and Kinetics
Edited by Danail Bonchev and Dennis H. Rouvray

This book is part of a series. The publisher will accept continuation orders which may
be cancelled at any time and which provide for the automatic billing and shipping of
each title in the series upon publication. Please write for details.

CHEMICAL GRAPH THEORY
Introduction and Fundamentals

Edited by

Danail Bonchev
Department of Physical Chemistry, Higher Institute of Chemical Technology, Burgas, Bulgaria

and

Dennis H. Rouvray
Department of Chemistry, University of Georgia Athens, Georgia, USA

CRC Press
Taylor & Francis Group
Boca Raton London New York

CRC Press is an imprint of the
Taylor & Francis Group, an **informa** business

First published 1991 by Gordon and Breach Science Publishers

Published 2019 by CRC Press
Taylor & Francis Group
6000 Broken Sound Parkway NW, Suite 300
Boca Raton, FL 33487-2742

© 1991 by Taylor & Francis Group, LLC
CRC Press is an imprint of Taylor & Francis Group, an Informa business

First issued in paperback 2019

No claim to original U.S. Government works

ISBN-13: 978-0-367-45070-0 (pbk)
ISBN-13: 978-0-85626-454-2 (hbk)

Visit the Taylor & Francis Web site at
http://www.taylorandfrancis.com

and the CRC Press Web site at
http://www.crcpress.com

Library of Congress Cataloging-in-Publication Data
Chemical graph theory : introduction and fundamentals / edited by D.
Bonchev and D. Rouvray.
    p. cm. -- (Mathematical chemistry series ; v. 1)
  Includes bibliographical references and references.
  ISBN 0-85626-454-7
    1. Chemistry--Mathematics. 2. Graph theory. I. Bonchev, Danail.
II. Rouvray, D. H. III. Series.
QD39.3.G73C48 1990
540'.15115--dc20

                                                                    90-42649
                                                                        CIP

CONTENTS

v

INTRODUCTION TO THE SERIES

The mathematization of chemistry has a long and colorful history extending back well over two centuries. At any period in the development of chemistry the extent of the mathematization process roughly parallels the progress of chemistry as a whole. Thus, in 1786 the German philosopher Immanuel Kant observed[1] that the chemistry of his day could not qualify as one of the natural sciences because of its insufficient degree of mathematization. It was not until almost a century later that the process really began to take hold. In 1874 one of the great pioneers of chemical structure theory, Alexander Crum Brown (1838–1922) prophesied[2] that "...chemistry will become a branch of applied mathematics; but it will not cease to be an experimental science. Mathematics may enable us retrospectively to justify results obtained by experiment, may point out useful lines of research and even sometimes predict entirely novel discoveries. We do not know when the change will take place, or whether it will be gradual or sudden...." This prophecy was soon to be fulfilled. Indeed, even before these words were uttered, combinatorial methods were being employed for the enumeration of isomeric species.[3] During Crum Brown's lifetime algebraic equations were used to predict the properties of substances, calculus was employed in the description of thermodynamic and kinetic behavior of chemical systems, and graph theory was adapted for the structural characterization of molecular species.

In the present century the applications of mathematics have come thick and fast. The advent of quantum chemistry in the 1920s brought in its wake a host of mathematical disciplines that chemists felt obliged to master. These included several areas of linear algebra, such as matrix theory and group theory, as well as calculus. Group theory has now become so widely accepted by chemists that it is now used routinely in areas such as crystallography and molecular structure analysis. Graph theory seems to be following in the footsteps of group theory and is currently being exploited in a wide range of applications involving the classification, systematization, enumeration and design of systems of chemical interest. Topology has found important applications in areas as diverse as the characterization of potential energy surfaces, the discussion of chirality, and the description of catenated and knotted molecular species. Information theory has yielded valuable insights into the nature of thermodynamic processes and the origin of life. The contemporary fascination with dissipative systems, fractal phenomena and chaotic behavior has again introduced new mathematics, such as catastrophe theory and fractal geometry, to the chemist.

All of these and numerous other applications of mathematics that have been made in the chemical domain have brought us to a point where we consider it may now be fairly said that mathematics plays an indispensible role in modern chemistry. Because of the burgeoning use of mathematics by chemists and the current feeling that mathematics is opening up some very exciting new directions to explore, we believe that the 1990s represent a particularly auspicious time to present a comprehensive treatment of the manifold applications of mathematics to chemistry. We were persuaded to undertake this somewhat awesome task after much reflection and eventually decided to publish our material in a series of volumes, each of which is to be devoted to a discussion of the applications of a specific branch of mathematics. The title of our series, *Mathematical Chemistry*, was chosen to reflect as accurately as possible the proposed contents. The term "mathematical chemistry" was coined in the early 1980s to designate the field that concerns itself with the novel and nontrivial application of mathematics to chemistry. Following the usual practice in this area, we shall interpret chemistry very broadly to include not only the traditional disciplines of inorganic, organic and physical chemistry but also its hybrid offspring such as chemical physics and biochemistry.

It is anticipated that each of the volumes in our series will contain five to six separate chapters, each of which will be authored by a leading expert in the respective field. Whenever it is evident that one such volume is insufficient to do justice to a wealth of subject matter, additional volumes devoted to applications of the same branch of mathematics will be published. In this way it is hoped that our coverage will indeed be comprehensive and reflect significant developments made up to the end of the twentieth century. Our aim will be not only to provide a background survey of the various areas we cover but also to discuss important current issues and problems, and perhaps point to some of the major trends that might be reasonably expected in mathematical chemistry in the early part of the new millennium. In the first few volumes of our series we propose to examine the applications to chemistry of graph theory, group theory, topology, combinatorics, information theory and artificial intelligence.

It may be of interest to observe here that mathematical chemists have often applied and even sought after branches of mathematics that have tended to be overlooked by the chemical community at large. This is not to imply that the mathematics itself is necessarily new — in fact, it may be quite old. What is new is the application to chemistry; this is why the word novel was employed in our earlier definition of mathematical chemistry. The thrill of discovering and developing some novel application in this sense has been an important source of motivation for many mathematical chemists. The other adjective used in our definition of mathematical chemistry, i.e. nontrivial, is also worthy of brief comment. To yield profitable new insights, the mathematics exploited in a chemical context usually needs to be of at least a reasonably high level. In an endeavor to maintain a uniformly high level, we shall seek to ensure that all of the contributions to our volumes are written by researchers at the forefront of their respective disciplines. As a consequence, the contents of our various volumes are likely to appeal to a fairly sophisticated audience: bright undergraduate and postgraduate students, researchers operating at the tertiary level in academia, industry or government service, and perhaps even to newcomers to the area desirous of experiencing an invigorating excursion through the realms of mathematical chemistry. Overall, we hope that our series will provide a valuable resource for scientists and mathematicians seeking an authoritative and detailed account of mathematical techniques to chemistry.

In conclusion, we would like to take this opportunity of thanking all our authors, both those who have contributed chapters so far and those who have agreed to submit contributions for forthcoming volumes. We should also like to thank our publisher, Gordon and Breach, for supporting us in this project and bringing out what we feel is a timely and worthwhile addition to the literature on mathematical chemistry. Finally, it is our sincere hope that the material to be presented in our series will find resonance with our readership and afford many hours of enjoyable and stimulating reading.

Danail Bonchev and
Dennis H. Rouvray

1. I. Kant, *Metaphysiche Anfangsgründe der Naturwissenschaft*, Hartknoch Verlag, Riga, 1786.
2. A. Crum Brown, Rept. Brit. Assoc. Sci., 45–50, 1874.
3. F.M. Flavitsky, *J. Russ. Chem. Soc.*, 3, 160, 1871.

PREFACE

The present work is the first of the volumes to appear in our new series entitled *Mathematical Chemistry*. As such, it represents the initial opus in what we anticipate will become an ongoing series of works devoted to discussion of the major applications of mathematics to chemistry. In our opening foray into the domain of mathematical chemistry we have elected to focus on the chemical applications to graph theory. This volume is intended to provide a useful introduction to the area by treating the fundamentals of the subject and some of its more important applications. Graph theory was chosen as our theme because its story typifies what is happening with many of the newer applications of mathematics in the chemical context. The applications and the number of publications deriving from chemical graph theory have grown enormously, starting from a mere trickle two decades ago and developing into the torrent that confronts us today. Graph theory has been employed in the classification, enumeration, systematization and design of many different kinds of system of chemical interest, ranging from individual molecules to complex reaction networks. Its potential for generating new concepts based on the structure of chemical systems and its capacity to reveal regularities in chemical data sets have rendered it a very valuable tool in the chemist's current arsenal. Moreover, because of the great generality of graph-theoretical methods, analogies between seemingly unrelated chemical systems and problems can often be discerned. It is our aim here to introduce the reader to some of the exciting possibilities that graph theory has to offer the chemist.

In Chapter 1 Rouvray surveys the development of chemical graph theory from a historical perspective, tracing the subject from its origins over 200 years ago down to the present day. Chapter 2 by Polansky provides an outline of the basic ideas and mathematical formalism of graph theory. Starting from the concept of the graph, Polansky proceeds to discuss such chemically relevant notations such as connectedness, graph matrix representations, metric properties, symmetry and operations on graphs. Having been thus primed, the reader is then led in Chapter 3 by Goodson into a discussion on the role of graphs in chemical nomenclature. This topic, which is all too often neglected by the chemist, is worthy of serious treatment because it has important implications for the storage and retrieval of chemical information. An intriguing discussion then follows on the relevance of graph-theoretical polynomials in Chapter 4 by Gutman.

Many specific types of polynomial, such as the characteristic and matching polynomials, are introduced and their uses described.

Chapter 5 by Balaban deals with the methodology of isomer enumeration, a field that continues to challenge us with many fascinating problems. Balaban's discussion embraces not only the traditional techniques adopted for enumeration of chemical species, such as the now classic Pólya method, but also delves into more recent approaches developed by workers such as Ruch and De Bruijn, Harary and Palmer. Our final Chapter by Trinajstić elucidates the interplay between graph theory and molecular orbital theory from the standpoint of spectral graph theory, and highlights in particular the concept of topological resonance in molecular species.

To conclude we should like to acknowledge a profound debt of gratitude to our many colleagues—too numerous to mention here by name—currently working in the field of chemical graph theory. Without their continual encouragement and support our own involvement in this field over the past two decades would not have been nearly as enjoyable and fulfilling. We can only hope that some of the stimulation and excitement that we have experienced will be shared by our readers. If this proves to be the case and eventually leads to their own contribution to the field we shall be well pleased.

Danail Bonchev and
Dennis H. Rouvray

Chapter 1

THE ORIGINS OF CHEMICAL GRAPH THEORY

Dennis H. Rouvray

Department of Chemistry, University of Georgia, Athens, GA 30602, USA

1.1 Introduction

It has frequently been remarked that mathematics is a more effective tool in the natural sciences than might be reasonably expected [1]. At first sight, the evident power of mathematics may indeed seem surprising, though more mature reflection on this theme leads us to anticipate the validity of mathematics in the description of the real world. Pure mathematics is founded on sets of axiomatic systems that form the basis for hypothetico-deductive theories of relations. Although in general not derived from observation of the real world, mathematical axioms do not normally violate observable phenomena. The natural sciences, however, which also employ hypothetico-deductive constructs, are tied to concrete

1

interpretation. The axioms upon which the sciences are built (the so-called laws of nature) must accord with observations of the real world. Since reality is constituted in the interaction of consciousness with an environment, any conceptual scheme of reality automatically reflects the processes occurring in the consciousness and the environment [2]. As products of the mind, both mathematical and scientific axioms will have a structure imposed on them [3] that is determined by the neural networks constituting the human brain [4]. The situation has been summed up by Weyl [5], who stated that "It would be folly to expect cognition to reveal to intuition some secret essence of things hidden behind what is manifestly given by intuition."

Viewed in this light, the various kinds of mathematics that have flourished in the past, and that might conceivably be developed in the future, can be regarded as potential starting points for construction of all the different possible sciences. The scientist thus plays the key role of establishing isomorphic relations between areas of mathematics and branches of science. This is usually accomplished by selecting appropriate structures from the mathematical storehouse and identifying them with scientific concepts. Relationships which are verified in the one domain then become of immediate applicability in the other domain. Results obtained or insights gained in the one system can thus be assumed to be directly transferable to the other. A classic example of this concordance of mathematics and science forms the subject matter of this chapter. We shall examine here the manifold interactions of the mathematical discipline of graph theory with the science of chemistry. In particular, we shall be tracing the origins of the interaction and focus on the early historical development of the field which has become widely known today as chemical graph theory.

Graph theory itself has a long and colorful history; the few words we now devote to this topic form a convenient departure point for the rest of our narrative. Graph theory is one of the few branches of mathematics that may be said to have a precise starting date [6]. In 1736, Euler [7] solved a celebrated problem, known as the Königsberg bridges problem. The question had been posed whether it was possible to walk over all the seven bridges spanning the river Pregel in Königsberg just once without retracing one's footsteps. Euler reduced the question to a graph-theoretical problem, and found an ingenious solution [7]. Euler's solution marked not only the introduction of the discipline of graph theory per se, but also the first application of the discipline to a specific problem. Since its inception, graph theory has been exploited for the solution of numerous practical problems, and today still retains an applied character. In the early days, very important strides were made in the development of graph theory by the investigation of some very concrete problems, e.g. Kirchhoff's study of electrical circuits [8], and Cayley's attempts to enumerate chemical isomers [9]. Further details on the history of graph theory may be obtained from the monograph by Biggs et al. [10]; two recent papers by Wilson [11,12] have also discussed the subject.

Before delving into those aspects of graph theory which relate specifically to chemistry, it is appropriate to make some comment on the term graph itself and its origins. We feel that it is not sufficiently widely known either by mathematicians or chemists that the term is of chemical origin. Although the word graph was first introduced into the literature by the mathematician Sylvester [13], it was derived by him from a contemporary chemical term. At the time, the chemical structure of a molecule was described as the "graphical notation" of a molecule. Sylvester felt that the word graph would be a convenient abbreviation for this chemical term. It was, of course, unfortunate that the word graph had already been applied in another context, namely, to describe cartesian data plots. However, the terminology has now become so well established that it is too late to consider making changes. It is therefore necessary to live with the fact that the word graph is used to describe two different concepts in mathematics, which are in no way related to one another.

1.2 The First Use of Chemical Graphs

Much of the current panorama of chemical theory has been erected on foundations that are essentially graph-theoretical in nature. Chemical graphs are now being used for many different purposes in all the major branches of chemistry. The present widespread usage of the chemical graph renders the origins of the earliest implicit application of graph theory of some considerable interest. Chemical graphs were first introduced in the latter half of the eighteenth century. To understand the need for them at that time and the circumstances of their introduction into the chemical literature, it will be necessary to say something about the prevailing attitudes in eighteenth century chemistry. Chemical thinking in the eighteenth century was steeped in Newtonian ideas, especially those pertaining to the internal structure of matter and the short-range forces existing between particles. In 1687 Newton himself had stated [14] that all natural phenomema depend "upon certain forces by which the particles of bodies, by some causes hithero unknown, are either mutually impelled towards one another, and cohere in regular figures, or are repelled and recede from one another."

It is perhaps not surprising, therefore, that contemporary chemists sought for explanations of chemical phenomena in terms of gravitational interactions. The typical attitude to chemical forces current in the eighteenth century is clearly revealed in the highly influential *Dictionary of Chemistry*, published by Macquer [15] in 1766. On the topic of gravity, he wrote: "we must examine what effects are produced by the gravity of bodies in their combinations and decompositions, that is, in all chemical operations. It is most important ... for the general theory of chemistry, but cannot be well treated but by the help of mathematics. In this point these two sciences, which appear so remote from each other, meet. A man sufficiently able in both might ... lay the foundation for a new physico-mathematical science." Even as these prophetic words were being written, the foundations of chemical graph theory had already been laid. The notion that agglomerates of atoms held together by mutually attractive forces could be represented in three-dimensional space by appropriate diagrams had been adumbrated several years before in the works of Boscovich [16] and Lomonosov [17].

The first chemical graphs, clearly recognizable as such, were drawn by the Scottish chemist, William Cullen. In 1758 Cullen started using so-called "affinity diagrams" in his lectures to represent the supposed forces, existing between pairs of molecules undergoing various chemical reactions. Unfortunately, these diagrams were used solely for illustrating his chemistry lecture notes, and none were ever published [18]. Similar diagrams to those of Cullen's were later published by Black, who falsely claimed to have invented them [19]; toward the end of the eighteenth century, such diagrams became commonplace in British chemistry textbooks of the period. Reproductions of two surviving diagrams, due to Cullen, are shown in Figure 1.

The numbers (or, in some cases, symbols) appearing between pairs of reacting substances represent the magnitude of the gravitational attraction existing between the substances. It should be mentioned in passing that these numbers have no physical basis whatsoever and thus express no more than a totally fictitious quantification of the imagined forces acting between the substances concerned.

It was probably after seeing such diagrams that the Irish chemist, William Higgins, had the inspiration of representing the forces between the constituent components, i.e. the atoms, of molecules. In a book

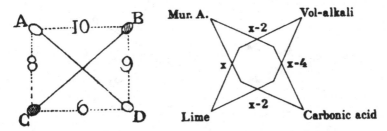

Figure 1. Examples of the first chemical graphs used by Cullen and Black in 1758 to represent the interactions of chemical substances. The supposed forces between pairs of substances are indicated either in terms of numbers or symbols.

published in 1789 [20], Higgins used a series of diagrams similar to those of Cullen, to portray a number of different individual molecules. Examples of some of his diagrams, depicting the five oxides of nitrogen, are reproduced in Figure 2.

In these diagrams the nitrogen atom is always represented by the symbol P (standing for phlogisticated air) whereas the oxygen atoms are denoted by the symbols a through e. Like Cullen before him, Higgins inserted arbitrary numbers between the various pairs of atoms in a vain attempt to quantify the force of attraction between them. It is important to emphasize here that the lines joining pairs of atoms are not to be interpreted as chemical bonds in the modern sense. The concept of the chemical bond was developed only some three quarters of a century later. The spatial arrangement of atoms was also not understood at the time of Higgins, and so all of his representations are two-dimensional. Moreover, in all of the diagrams in Figure 2, atoms are portrayed in topologically incorrect positions. In spite of these evident drawbacks, however, the insights of Higgins were quite remarkable for his time.

1.3 The Emergence of Structure Theory

It was not until the beginning of the nineteenth century that the first serious attempts were made to study the spatial arrangement of the atoms in molecules. Pioneering work in this area by Dalton and Wollaston led to a greatly improved understanding of the relationships between atoms in space and, incidentally, also gave rise to a new nomenclature for chemical

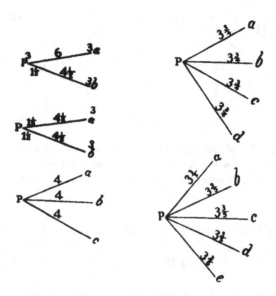

Figure 2. Reproduction of some of the first chemical graphs used by Higgins in 1789 to represent individual chemical species. Depicted here are the five oxides of nitrogen. The numbers indicate the supposed forces between pairs of atoms.

species. Both Dalton and Wollaston made frequent use of models, the latter averring [22] that "when our views are sufficiently extended ... to reason with precision concerning the proportions of elementary atoms ... we shall be obliged to acquire a geometrical conception of their relative arrangements in all three dimensions of solid extension." Dalton had constructed a large number of models of the ball-and-stick variety [23]. He envisioned atoms as tiny, hard spheres that were held together in molecules by forces which he modeled by means of sticks. In depicting molecules, Dalton represented the atoms by circles containing a symbol for each specific type of atom, a device he adopted because the circle is the two-dimensional analogue of the sphere. Molecules were drawn as ensembles of touching circles and classified according to the number of atoms they contained. Several examples of Dalton's molecules, reproduced from his famous textbook [24], are shown in Figure 3.

Although Dalton had constructed many of the models now familiar

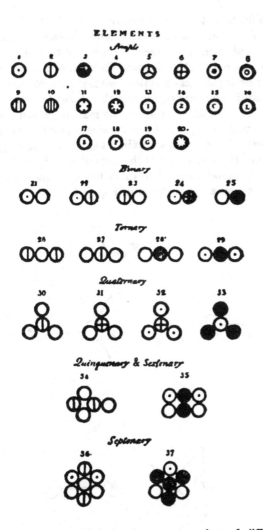

Figure 3.   Reproductions of Dalton's representation of different classes of molecules, taken from his celebrated book *New System of Chemical Philosophy* (1808).

to modern stereochemistry around the turn of the nineteenth century, it was not until the mid part of the century that spatial notions began to have a significant impact on chemistry as a whole. The person chiefly responsible for raising the status of three-dimensional thinking was an architect turned chemist: August Kekulé. Until Kekulé's time, it was common to refer only to the *chemical* positions of the atoms in molecules, that is to their topological positions within the molecule. Kekulé broke with this tradition by considering also the *physical* positions of the atoms, i.e. their relative orientation in space. By making extensive use of a variety of models, including models of the tetrahedral carbon atom, Kekulé was able to classify the structures within a number of homologous and other organic series [25], and visualize the formation of both double and triple bonds [26]. A reproduction of Kekulé's two-dimensional depiction of the tetrahedral carbon atom is shown in Figure 4.

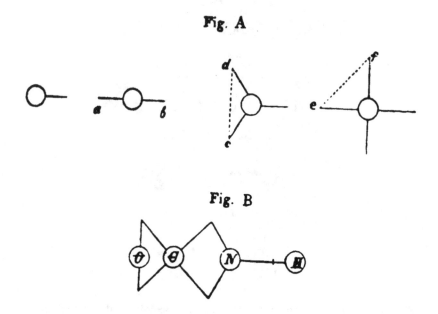

Figure 4.   Reproduction of Kekulé's earliest attempt to depict the tetrahedral carbon atom dating from 1867.

The tetrahedral carbon atom model was used to great effect in 1874 independently by van't Hoff and Le Bel to interpret the isomerism extant in a wide range of organic species [27].

The first cyclic chemical species to be studied in detail was the benzene ring. Although even Dalton had constructed hexagons of atoms [23], it was not until 1854 that Laurent published the first illustration of the benzene ring [28]. A reproduction of this illustration is presented in Figure 5(a). In 1861 Loschmidt [29] published graphic representations for the structural formulas of over 350 different molecules. In this system the hydrogen atoms were represented by small circles, carbon atoms by larger circles, and the benzene ring by one very large circle. Loschmidt's representation of the benzene molecule is reproduced in Figure 5(b). His work was of great significance since it depicted for the first time both double and triple bonds as well as offering a representation of the benzene nucleus as a single ring. The now famous structural formula for benzene published by Kekulé [30] in 1866 is shown in Figure 5(c). Letters were placed at the vertices of his hexagon to assist in the enumeration of the isomers formed when the hydrogen atoms were substituted by various heteroatoms. Shortly after the appearance of Kekulé's paper [30], some surprisingly modern looking representations of the naphthalene molecule [31] and anthracene molecule [32] were published by Graebe.

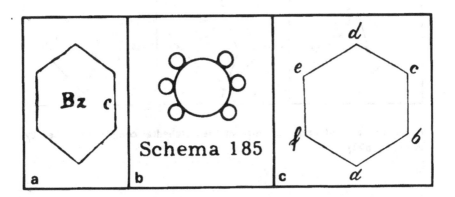

Figure 5. Reproductions of the three earliest representations of the benzene ring by (a) Laurent (1854), (b) Loschmidt (1861), and (c) Kekulé (1866).

After the introduction of the tetrahedral carbon atom concept into organic chemistry, various polyhedral structures were postulated in the inorganic domain. Work in the field of coordination chemistry, which was developed virtually single-handedly by Werner [33], led to the idea that the number of atoms located around some central atom (the coordination number of the atom) could vary depending on the nature of the atom and ligands involved. Werner made use of the octahedron to represent the structure of most of the complexes he studied [34]. Three examples of his use of octahedra to illustrate the number of isomers formed in octahedral ML_6 compounds are shown in Figure 6. Chemical polyhedra having more vertices than octahedra first made their appearance in the twentieth century. The earliest use of the icosahedron arose from studies on the chemistry of boron compounds. The icosahedron was first used for the representation of boron carbide by Clark and Hoard [35] in 1943. Later work by Lipscomb and coworkers [36] introduced into chemistry a large number of three-dimensional boron structures, many of which were either complete icosahedra or icosahedral fragments. The dodecahedron made its debut in 1958 when Pauling [37] used it in his discussion of the structure of water and hydrated clathrate compounds. Truncated icosahedra, such as buckminsterfullerene, have recently been proposed [38] for C_{60} and other carbon atom clusters.

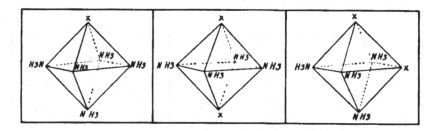

Figure 6.   Reproductions of three of the octahedral complexes studied by Werner (1893).

1.4 The Concept of Valence

The view that each different chemical substance may be associated with a fixed molecular structure, and that this structure could be elucidated by chemical means, was of central importance in the development of modern chemical thinking. Indeed, as Hein [26] has commented, this construct "may be the most fruitful conceptual scheme in all of the history of science". The scheme is usually designated nowadays as structure theory, following upon introduction of the term 'molecular structure' by Butlerov [39] in 1861. The evolution of the concept, with evaluations of the pioneering contributions of workers such as Butlerov, Couper, Crum Brown, and Kekulé, has been discussed by Larder [40] and Rocke [41], and in the monographs of Russell [42] and Kuznetsov [43]. Difficulties arising from the chemical analysis of compounds and in the visualization of chemical structures made the germination of structure theory a long and convoluted one. This is evidenced in Figure 7, which presents a pictorial illustration of several of the early attempts to formulate a structure for the acetic acid molecule. In going from Döbereiner's straight chain representation of 1822 through Couper's complex structure of 1858, only Williamson's formula succeeded in showing the correct stoichiometry. Couper's formula comes closest to depicting the actual structure.

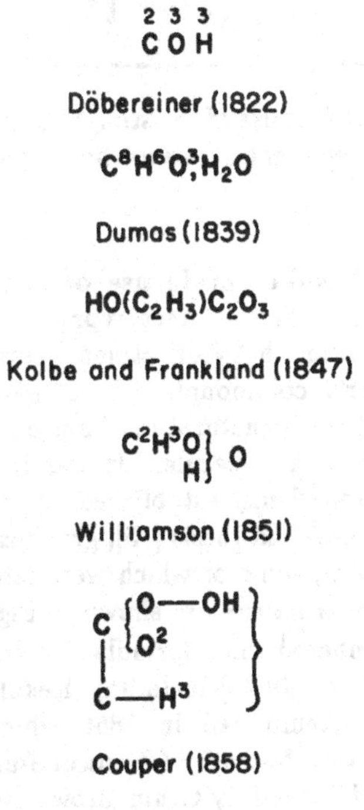

$$\overset{2\ 3\ 3}{C\ O\ H}$$

Döbereiner (1822)

$$C^8H^6O_3^{\,}\!,H_2O$$

Dumas (1839)

$$HO(C_2H_3)C_2O_3$$

Kolbe and Frankland (1847)

$$\left.\begin{matrix}C^2H^3O\\H\end{matrix}\right\}\ O$$

Williamson (1851)

$$\overset{}{\underset{\displaystyle C}{\overset{\displaystyle C}{\rule{0.4pt}{18pt}}}}\left\{\begin{matrix}O\text{---}OH\\O^2\\\\-H^3\end{matrix}\right\}$$

Couper (1858)

Figure 7. Illustration of some of the earliest attempts to formulate and depict the structure of the acetic acid molecule.

Couper's formulation is of especial importance in that it contains the first depiction of a chemical bond by means of a straight line connecting two bonded atoms. It is thus to Couper that the credit goes for the first use of the graphical edge to represent a chemical bond [44]; other instances of his use of this device are reproduced in Figure 8.

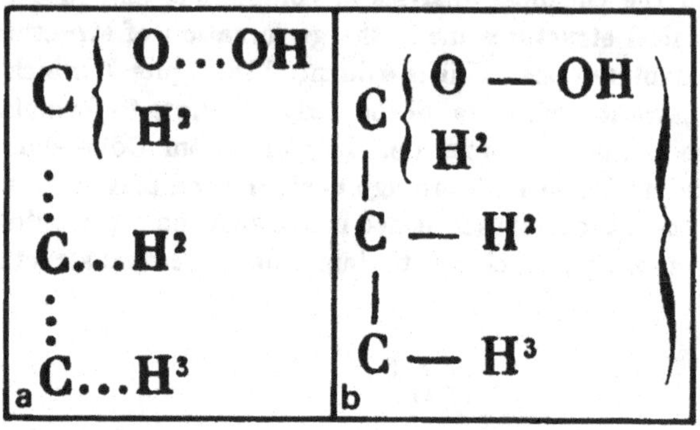

Figure 8. Reproduction of the first use of the straight and dotted lines drawn between a pair of atoms to represent a valence bond, published by Couper (1858).

Although Higgins had earlier made use of straight lines in his depictions of chemical structures, his lines represented no more than vague, interatomic forces extant between atoms. During the first half of the 1860s it became fairly commonplace to attempt to understand chemical species in terms of the structural relations existing between the various pairs of atoms within a molecule. It was in this period that the structural formula became firmly established. Thus, in 1861, Crum Brown presented, in an unpublished thesis [45], graphical representations of a wide variety of molecules, some of which were later published [46]. Reproductions of two of his structures are shown in Figure 9.

By 1865, the six-membered ring formula for benzene had been postulated by various workers [28,29] including Kekulé [47]. The new formulas became widely disseminated in 1866 when Frankland [48] published his celebrated *Lecture Notes for Chemical Students*, in which he made use of the notation advocated by Crum Brown [46]. Within a year or so, the drawing of circles around the individual atoms was discontinued, and, in virtually all respects, modern structural formulas were launched and rapidly came into general use.

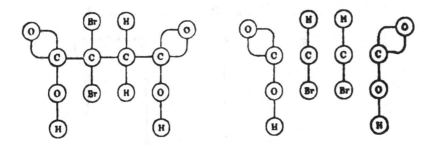

Figure 9. Reproductions of the graphical formulas employed by Crum Brown (1864) to represent the topological positions of atoms in a variety of molecules.

Frankland had championed the idea that individual atoms were always associated with a fixed valence, and drew atoms with a fixed number of attached valence bonds [48], as illustrated in Figure 10.

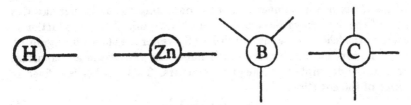

Figure 10. Reproductions of atoms showing fixed valence bonds as published by Frankland (1866) in his celebrated book *Lecture Notes for Chemical Students*.

This idea led to the introduction of a number of essentially graph-theoretical results based on chemical structures. As far back as 1843, Laurent [49] and Gerhardt [50] had established that the number of atoms of odd valence in a molecule was always even. Observations on various homologous series led Kekulé to the representation of such series by general formulas; for the alkane series he obtained the formula C_nH_{2n+2} [51]. Two mathematicians used structural formulas for the construction of chemical graphs. Cayley [52] constructed so-called 'kenograms' which were alkane tree graphs; these were used in the enumeration of the alkane structural isomers, a topic we discuss below. Examples of Cayley's trees are shown in Figure 11.

$n = 1.$ $n = 2.$ $n = 3.$ $\dot{n} = 4.$

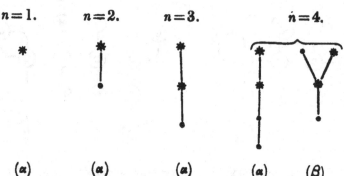

(α)   (α)   (α)   (α)   (β)

Figure 11. Reproductions of tree graphs called "kenograms" as used by Cayley (1874) in his enumeration of the alkyl radicals.

The mathematician Sylvester [53] investigated the conditions for the existence of chemical graphs and concluded that 'if the difference between every two letters of an algebraically existent graph be raised to the power whose index is the number of bonds connecting them the permutation sum of the product of those powers must not vanish.' A reproduction of some of the 'chemicographs' studied by Sylvester are shown in Figure 12.

The concept of the cyclomatic number has also played an important part in the description of chemical structure. This number is defined in terms of the equation:

$$\mu = n_e - n_v + 1, \tag{1}$$

where the cyclomatic number, μ, is the number of independent cycles in the graph, and n_e and n_v are, respectively, the numbers of edges and vertices in the graph. The mathematician Clifford [54] first demonstrated that a hydrocarbon of the form C_nH_{2n+2} could not possess any cycles, and that a hydrocarbon with the general formula $C_nH_{2n+2-2x}$ would contain x cycles. For present purposes a double bond counts as a single cycle (i.e. as a lune) in the graph. The use of the cyclomatic number was extended by Lodge [54] who obtained closed formulas for the number of cycles in a variety of organic and inorganic species. He was able to prove, for instance, that the number of cycles contained in the general covalent molecule $K_rL_sM_t\dots$ would assume the form:

$$\mu = \frac{1}{2}[r(k-2) + s(l-2) + t(m-2) + \dots] + 1, \tag{2}$$

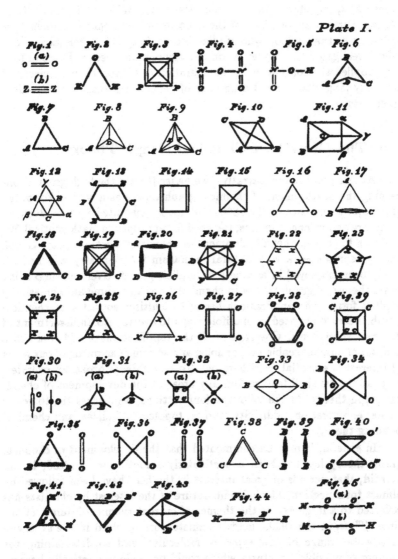

Figure 12.   Reproductions of the chemical graphs studied by Sylvester (1878). By lettering the vertices of the graphs, Sylvester was able to devise a prescription for the existence of different types of chemical graphs.

where k, l, m_i etc. represent respectively the valence of atoms K, L, M, etc. Further applications of the concept in the chemical domain were made by Kurnakov [55], who studied inorganic species, Frèrejacque [56], who investigated the condensation of large organic species; Balaban [57], who applied the concept in the enumeration of hydrocarbon species; and Rouvray [58], who derived a number of general expressions for covalent molecules.

1.5   The Growth of Chemical Graph Theory

At this point in our narrative, we pause to gain the high ground concerning the development of chemical graph theory as a whole. Following the major turbulence of the 1850s and 1860s, generated by the advent of structure and valence theories, the need for appropriate mathematical formalisms for the further development of chemistry became clear. Graph theory represented a very natural formalism for chemistry and had already been employed in a variety of implicit guises. Over the succeeding years the use of graph theory in chemistry was to assume an increasingly explicit form. The applications began to multiply so fast that chemical graph theory bifurcated in manifold ways to evolve into an assortment of different specialisms. Coverage of all the ramifications would result in a chapter of unmanageable proportions, and so some narrowing of our focus of interest is essential from here on. To maintain as broad an overview as possible, we shall feature four key areas of the development of chemical graph theory. In the following four sections, we discuss the topics of isomer enumeration, additivity studies, topological indices, and chemical bonding theory.

In general, it may be commented that the development of chemical graph theory has not been a particularly smooth one. Indeed, it can be said, that periods of great interest in the field have been followed by almost total neglect. The episodic nature of the interest by chemists has reflected to a large extent the themes of current chemical interest at the time. Thus, in the 1930s, many chemists were involved in the synthesis of a whole range of new types of molecules, and so determining the number of possible structures which could be made theoretically became of some significance. It was during the 1930s that isomer enumeration studies had their heyday. Chemical bonding theory had been a topic of perennial interest, though the topological nature of simple bonding theory began to be realized only in the late 1950s. As a consequence, in the period immediately following, an explosion occurred in the number of papers treating the subject of bonding theory from a graph-theoretical standpoint. In very recent times, in the 1970s and 1980s, chemical graph theory seems to have become broadly fashionable again, even among a fair number of mathematicians.

The current fashionability of our subject appears to date from the 1960s, when Balaban began publishing an important series of papers bearing the general title 'Chemical Graphs'. One measure of the interest in a subject is the number of papers published in that field. If we apply this test to chemical graph theory, it would seem that Balaban's pioneering work sparked a growing avalanche of interest. In Figure 13 a plot is shown of the annual number of papers published in the field for the years 1970-1986 based on listings from *Chemical Abstracts*.

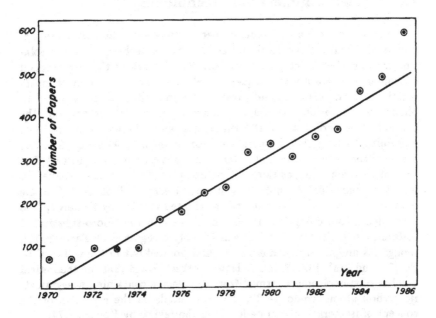

Figure 13. A plot showing the annual number of papers published in the area of chemical graph theory for the years 1970-1986. Note the annual growth rate of around 25% for this period.

The plot reveals that whereas the number was comparatively small (i.e. around 70) for the year 1970, it had increased to over 500 by 1986.

This represents a more or less steady growth rate of some 25% per annum for this period. At present, this growth shows no sign of abating, and so we may confidently predict a high level of interest in chemical graph theory for several years to come. Future reviews of the field are likely to focus on specialized themes rather than attempt the near impossible task of embracing the whole area in a single survey. Previous reviews which provide a reasonably comprehensive coverage are due to Rouvray [59], Rouvray and Balaban [60], and Balaban [61]. A two-volume monograph by Trinajstić [62] also offers an excellent introduction to the subject. Collections of research papers in the area of chemical graph theory have been edited by Balaban [63], King [64], and King and Rouvray [65].

1.6   Isomer Enumeration Techniques

The enumeration of isomeric species represents the first overt application of graph-theoretical and combinational techniques to the solution of a problem of chemical interest. The history of the development of isomer enumeration techniques reveals a strong interaction of mathematicians and chemists, and provides a textbook example of interdisiplinary collaboration. Although of much more ancient vintage [66], the concept of isomerism was first formally defined by Berzelius [67] in 1830. According to Berzelius, isomers were compounds, "possessing the same chemical constitution and molecular weight but differing properties." The earliest systematic exposition of the theory of structure is contained in Kekulé's famous *Lehrbuch der Organischen Chemie* [68] of 1861. The role of structural isomerism was further elucidated in 1862 by Butlerov, who succeeded in obtaining the number of isomers in the chloro-substituted methanes [69]. Stereoisomerism was initially recognized by Pasteur [70], though its ubiquity in chemistry was first pointed out in the works of Le Bel [71] and van't Hoff [72]. The latter worker showed that the theoretical number of stereoisomers formed from a molecule containing n asymmetric carbon atoms would be 2^n. Further details on the evolution of the concept of isomerism are to be found in the review by Rouvray [74].

Close on the heels of the first use of chemical graphs for the representation of individual molecules, came the realization that the same graphs could be of great value in the discussion of isomerism. As mentioned above, chemical graphs provide information only on the topological positions of atoms and not on their physical positions in the molecules. Such graphs are therefore well suited to the enumeration of structural isomers, but are not appropriate for the treatment of stereoisomers. This was clearly established even in the first use of chemical graphs for this purpose. In 1865, Crum Brown [75] applied his diagrams (see Figure 9) to a study of different kinds of isomers and was led to the conclusion that, although they could readily explain the isomerism among the members of homologous series such as the alkanes and alchohols, they could not account for the isomerism in the tartaric acids. Stereoisomerism could be effectively tackled only after the pioneering work of Le Bel [71] and van't Hoff [72] in the organic sphere, and of Werner [33] in the inorganic sphere.

The earliest use of a combinatorial technique to determine isomer counts appears to be that of Flavitsky [76] who studied the alcohols. By developing sets of recursion relations for the saturated alcohols, Flavitsky was able to derive formulas which yielded the number of primary, secondary, and tertiary alcohols for a given number of carbon atoms. Since he also demonstrated that the number of primary alcohols having n atoms is equal to the total number of all alcohols having $n - 1$ carbon atoms, he was able to derive a table of values for primary, secondary, tertiary, and the total number of alcohols. His table, which lists results for the first ten alcohols in the series is reproduced in Figure 14.

It is interesting to note that the isomer counts he gave accord with those based on modern computer computations [77]. The approach of Flavitsky was greatly extended in the 1930s when Henze and Blair [78] started developing recursion formulas for the members of several homologous series. Their results on the alcohols [78] greatly expanded on those of Flavitsky, and they went on to show the applicability of their method to many other series, including the aldehydes, alkyl halides, amines, ethers and organic acids [79,80]. Several further extensions of this general approach have been made in more recent years [73].

Generating functions were first employed in isomer enumeration work by the mathematician Cayley. In 1857, Cayley [81] had devised a method for enumerating rooted trees which centered on a generating function of the general form:

$$(1+x)^{-1}(1-x^2)^{-A_1}(1-x^3)^{-A_2}\ldots(1-x^n)^{-A_{n-1}} = 1+A_1x+A_2x^2\ldots \quad (3)$$

where x is a variable, n the number of vertices, and A_{n-1} the coefficient of x^{n-1}, which is equal to the number of rooted trees on n vertices. Using this function, he succeeded in 1874 in enumerating unrooted trees [52]

Эмпирическая фор-мула алкоголей.	Число изомеровъ алкоголя:			
	первичныхъ.	первичныхъ	вторичныхъ.	третичныхъ.
C_2H_6O	$U_2=1$	1	—	—
C_3H_8O	$U_3=2$	1	1	—
$C_4H_{10}O$	$U_4=4$	2	1	1
$C_5H_{12}O$	$U_5=8$	4	3	1
$C_6H_{14}O$	$U_6=17$	8	6	3
$C_7H_{16}O$	$U_7=39$	17	15	7
$C_8H_{18}O$	$U_8=89$	39	33	17
$C_9H_{20}O$	$U_9=211$	89	82	40
$C_{10}H_{22}O$	507	211	194	102

Figure 14. Reproduction of the first table of isomer counts enumerated by Flavitsky (1871). Moving from left to right, this table presents for each of the first ten members of the alcohol series the chemical formula, the total number of alcohols U_n, and isomer counts for the numbers of primary, secondary, and tertiary alcohols.

which correspond to alkane molecules (see Figure 11). He obtained isomer counts for alkanes up to the thirteenth member of the series, though his results for the latter two were shown to be erroneous. By 1875 Cayley had published a lengthy paper [9] wherein he considered trees representing molecules in which the maximum valence of the atoms was four. He thus considered not only the alkanes, but also so-called 'boron trees' (maximal valence three), and 'oxygen trees' (maximal valence two). He also presented isomer counts for the alkyl radicals up to the thirteenth member [82]. The errors in Cayley's work were corrected in 1889 when Hermann [83] published accurate counts for the alkanes. Hermann based his enumerations on a classification of trees into types determined by the number of branches attached to the main chain [83].

The roles of molecular configuration and symmetry began to be seriously explored in the 1920s. In 1929, Lunn and Senior (a chemist and a mathematician, respectively) published a treatise [84] in which they stressed the importance of permutation group theory in the characterization of molecular configurations. They developed a closed mathematical relationship, applicable for the enumeration of substitutional isomers, which took the form:

$$N_G(p) = \frac{1}{g} \sum n_t K(p, t), \qquad (4)$$

where $N_G(p)$ is the isomer count for a set of molecules associated with the permutation group G within some class p, g is the order of group G, n_t is the number of symmetry operations of type t that may be performed on the molecule, and $K(p, t)$ is the number of invariant configurations left in p when the operation t is carried out. Lunn and Senior [87] also presented extensive tables documenting $K(p, t)$ values for various molecular configurations. In the purely mathematical domain, a seminal work by Redfield [85] had been published in 1927 that much later was seen to have set forth constructs that subsumed those of Lunn and Senior and many subsequent workers. This important paper was all but ignored upon its publication and achieved celebrity status only many decades later [86].

The concepts of generating functions, symmetry classes of molecules, and permutation groups were all incorporated within the framework of a very powerful mathematical theorem published in 1937. This theorem, known today as the Enumeration Theorem of Pólya [87], has been extensively used to enumerate a wide variety of different types of isomers. The enumerations are accomplished by taking into account the different symmetry operations that can be performed on the chemical graphs using only the proper rotation axes. The symmetry of a molecule is expressed as a so-called cycle index, a polynomial written in terms of variables characterizing the molecular symmetry. The cycle index is formally defined as:

$$C(G) = \frac{1}{g} \sum{}' h_{j_1 j_2 \cdots j_p} f_1^{j^1} f_2^{j^2} \cdots f_p^{j_p}, \qquad (5)$$

where $C(G)$ is the cycle index corresponding to a group G of order g, p is the number of vertices permuted, the f_p are the variables, and $h_{j_1} \cdots_{j_p}$ is the number of permutations of G consisting of j_1 cycles of order one up to j_p cycles of order p. The prime on the summation signifies that the condition:

$$\sum_{c=1}^{p} h_{j_c} = P \tag{6}$$

must hold, where P is the total number of permutations. A figure counting series is substituted into each of the terms in the cycle index to yield a configuration counting series. The isomer counts are then obtained directly as the coefficients of this latter series.

To date, very extensive use has been made of Pólya's Theorem for the enumeration of many different types of isomers. Examples include isotopic isomers [88], cycloalkanes [89], substituted porphyrins [90], polycyclic aromatic hydrocarbons [91], and inorganic complexes [92]. Further details on applications of the theorem in the chemical context are to be found in the review of Rouvray [73], in the monographs of Balaban [63] and Trinajstić [62], and in a chapter by Balaban elsewhere in this volume. More recently developed techniques for isomer enumeration, such as the use of power groups by De Bruijn [93], the double coset formalism of Ruch [94], or the reformation of Pólya's Theorem using generalized wreath products by Balasubramanian [95], are discussed in some detail in the chapter by Balaban alluded to above in this volume of *Mathematical Chemistry*.

1.7   Early Additivity Studies

The earliest use of a graph-theoretical approach for the elucidation of physicochemical phenomena dates from the first half of the nineteenth century. It was toward the end of this period that several workers began exploiting essentially graph-theoretical concepts in the investigation of the so-called additive properties of molecules. An additive property is defined as one that can be determined by summing the contributions to that property from each of the component parts of a molecule. Although, strictly speaking, there are no additive properties of molecules, many molecular properties do display a close approach to additive behavior. This is particularly true for the members of homologous series, which probably come closest to being additive. The concept of homologous series of compounds was introduced in 1845 by Gerhardt [96], who stated that such compounds have "similar properties and the composition offers certain analogies in the relative properties of the elements". He also coined the term homology to characterize the law of succession of the members of homologous series. The establishment of this concept was to play a key role in understanding additive behavior.

In 1839 Kopp [97] had begun investigations on the atomic volumes and boiling points of series of compounds, many of which were members of homologous series. By 1842 his studies had led him to whole sets of new and unsuspected relationships. Kopp's papers [98,99] are now viewed as containing the first statement of the additivity principle. Numerous experiments had revealed, for instance, that ethyl esters boil some 18° higher than their methyl counterparts, and that the substitution of a hydrogen atom by chlorine in organic species resulted in a rise of around 25°C in the boiling point. Based on such observations, Kopp developed algebraic formulations of the relationships existing between the members of different series of compounds [98,99]. His formulations were expressed both as algebraic equations and as matrices; examples of his formulations are reproduced in Figure 15. The sets of equations that he deduced were deemed to be valid for all the additive physicochemical properties of the various combinations of compounds included in his studies. In view of the low accuracy of the parameters he was using, it is remarkable that he was able to reach such sophisticated conclusions.

After Kopp's time, a host of workers began using the additivity principle in a variety of guises, initially in implicit form. The principle first received a formal statement in 1917 when Redgrove [100] presented the following definition: "Every physicochemical property which is additive in the broadest sense of the word, is a function of the number and sort of valency bonds or energy units of which the molecules of the bodies exhibiting the property in question may be regarded as complexes." To make his statement mathematically precise, he accompanied it by a so-called "proof" for the case of the general hydrocarbon, which was assigned the formula $C_nH_{2(n+1-x-2y-p)}$, where x and y represent respectively the number of double and triple bonds, and p is an integer. His work is important not only for the formal statement of the principle, but also because allowance is clearly made for every valence linkage in the molecule. The prescription he used to obtain an estimate of a physicochemical property was to evaluate the "sum of a series of products obtained by multiplying the number of each sort of bond by a quantity varying according to the bond, but always constant for the same bond." This general approach was adopted by Fajans [101] to calculate the energies of organic molecules, but quickly ran up against the problem of non-transferability of the bond energy terms from one molecule to another.

	A essigsaures		B ameisensaures		C benzoësaures
α Wasser . . .	709	—	467	—	?
β Aethyloxyd .	1243	—	1020	—	1794
γ Methyloxyd .	1012	—	?	—	1558.

Bezeichnen wir die specifischen Volume von A + α, A + β u. s. f. der Kürze halber durch diese Zeichen selbst, so ist innerhalb der Gränzen der Beobachtungsfehler:

$$(A + \alpha) - (A + \beta) = (B + \alpha) - (B + \beta)$$
$$(A + \beta) - (A + \gamma) = (C + \beta) - (C + \gamma)$$
$$(A + \alpha) - (B + \alpha) = (A + \beta) - (B + \beta)$$
$$(A + \beta) - (C + \beta) = (A + \gamma) - (C + \gamma)$$

und so fort.

Genau dieselbe Gesetzmäfsigkeit hat statt, wenn wir die beobachteten Siedepunkte der gedachten Verbindungen schematisch ordnen:

	A essigsaures		B ameisensaures		C benzoësaures
α Wasser . . .	120°	—	99°	—	239
β Aethyloxyd . .	74	—	53	—	209
γ Methyloxyd . .	58	—	?	—	198

Figure 15.　Reproduction of certain of the sets of equations and tables constructed by Kopp (1842) in his study of the additive properties of various series of molecules.　The upper table relates to specific volumes and the lower table to boiling points. In each case the properties of 1:1 mixtures are presented. Here A is acetic acid, B formic acid, C benzoic acid, α water, β diethylether, and γ dimethylether.

In an attempt to overcome this difficulty, Zahn [102] put forward the suggestion that only part of the energy of a molecule be ascribed to the bonds and that the rest be associated with adjacent pairs of bonds. This proposal was improved upon by Allen [103], who advocated taking into account the contributions from adjacent trios of bonds. For further details on the early development of these ideas, the reader is referred to the review by Somayajulu *et al* [104]. The first graph-theoretical formulation of the additivity principle was made by Smolenski [105] in 1964. The latter worker employed the basic equation:

$$P(G) = a_0 + \sum_{r=1}^{k} a_r H_r, \tag{7}$$

where $P(G)$ is some physicochemical property displayed by a set of molecules represented by the graph G, the a_r $(0 < r < k)$ are the constants determined experimentally from a small set of different types of molecules, H_r is the number of subgraphs of G of length r, and the summation extends over subgraphs of all lengths r. In the general formulation, allowance was made for the fact that the subgraphs of length r can represent differing components of the molecule. Such reasoning led to the more general equation:

$$P(G) = a_0 + \sum_{r=1}^{k} \sum_{b=1}^{\sigma_r} a_r^b H_r^b \tag{8}$$

where H_r^b represents the number of components of length r belonging to some equivalence class b, and the double summation is used to sum over both the k equivalence classes and the σ_r members of each of these classes.

A further generalization of earlier graph-theoretical formulations was achieved in 1973 when Gordon and Kennedy [106] introduced the concept of "graph-like states of matter." This concept was used to describe physicochemical properties of molecules that satisfied summations of the type:

$$P(G) = \sum_{i} \alpha_i T_i \tag{9}$$

where the α_i are coefficients that are either determined empirically or calculated by means of combinatorial techniques, and the T_i are graph-theoretical invariants which characterize the molecular graphs. Such invariants are nowadays referred to as topological indices. The T_i can assume any form and, for instance, in the example mentioned by the authors, might be the number of ways in which the graph of the molecule under consideration can be excized from some larger graph representing a skeleton of carbon atoms. Of course, we know that no physicochemical properties will satisfy equation (9) precisely, but those which do so to a fair approximation are said to belong to graph-like states of matter and are treated as additive properties of molecules.

1.8 The Introduction of Topological Indices

By their very nature, additivity studies have been restricted to properties that are more or less proportional to the molecular volume. In the case of hydrocarbons, the relationship of molecular volume to the number of carbon atoms present in the molecule has been examined by Calingaert and Hladky [107] and Kurtz and Lipkin [108]. These workers concluded that the molecular volume is directly proportional to the number of carbon atoms for the members of a variety of homologous series. The number of carbon atoms in a hydrocarbon molecule may be viewed as a graph-theoretical invariant for that molecule. It is equal to the number of vertices in the hydrogen-suppressed graph of the molecule. This graph invariant is recognized nowadays as the first topological index, the contemporary term used to designate a graph invariant that assumes the form of a scalar parameter. The earliest use of this index can be traced back to the work of Kopp [109], who summed the atoms of different kinds in molecules in an attempt to determine the specific volumes and densities of those molecules. Early examples of the use of this index for the prediction of the boiling points of alkane species are listed in the Table.

The trend to explore properties in which the constitutive nature of the property plays an increasingly important role became pronounced during the early part of the twentieth century [110]. This is clearly evidenced in the work of Sugden [111], who introduced the concept of the parachor in 1924. The parachor, which was designed to correlate physicochemical properties with molecular structure, was defined by the equation:

$$P(G) = \frac{m}{D-d}\,\gamma^{\frac{1}{4}}, \tag{10}$$

where m is the molecular mass, D and d represent respectively the densities of the liquid and vapor at a given temperature, and γ is

Author(s)	Year	Boiling Point Relationship
Kopp	1842	$\Delta t = 18$
Goldstein	1879	$\Delta t = 19 + 380/n(n+1)$
Mills	1884	$t = \alpha(n - \beta)/(1 + \gamma(n - \beta))$
Longinescu	1903	$t = n(100D)^{\frac{1}{3}}$
Young	1905	$\Delta t = 144.86/t^{0.0148t^{\frac{1}{2}}}$
Plummer	1916	$t = \alpha\log(\beta n + \gamma) + 70/2^n$
Egloff *et al*	1940	$t = 745.42\log(n + 4.4) - 416.31$

Table: Boiling point relationships for members of the normal alkane series. Here n is the number of carbon atoms in the molecule, t is the boiling point of the n-th member, Δt the difference in boiling point temperature between the n-th and $(n + 1)$-th member of the series, D is density, and α, β, and γ are constants.

the surface tension of the liquid at the same temperature. It was quickly established that the parachor provides a measure of molecular volume [111]. Thus, all the alkane isomers corresponding to a given member of the series are associated with roughly the same parachor because all of them have more or less the same size. In spite of this, efforts were made to apply the parameter to the elucidation of structural problems. Values of the parachor were determined for a wide variety of multiple bonds and ring systems [111]. The inevitable conclusion reached, however, was that the parachor is not appropriate for differentiating between the possible structures a molecule might adopt [112].

Numerous attempts were subsequently made to allow for structural variations rather than for pure molecular volume effects. We now make brief mention of several of the workers who contributed to this area. In 1929 Nekrassow [113,114] introduced the concept of "active structural elements" and made a variety of corrections for these in the calculation of the boiling points of hydrocarbon species. Taylor *et al* [115] devised formulas for determining a number of the properties of hydrocarbons and included contributions from the interactions occurring between adjacent component paths of the molecule. Mumford [116] examined the shape of the plots of different properties of normal alkanes against the carbon number index and pointed out a discontinuity in the region of the C_{16} member. The reason for the "discontinuity" was later explained by Rouvray and Pandey [117] and shown to be due to the fractal nature of alkane species. In his study of heats of formation, combustion, and vaporization, Laidler [118] developed a calculational system in which the numbers of atoms of various kinds in a molecule could be expressed in terms of the numbers of different kinds of bonds. Tatevski and Papulov [119] incorporated contributions for pairs of bonds removed through a C–C bond in their determination of physicochemical properties. This latter idea was further generalized by Somayajulu and Zwolinski [120], who elaborated a very general procedure allowing for many kinds of interactions in the calculation of the physicochemical properties of alkane species.

Apart from the use of the carbon number index, the first use of graph invariants for the correlation of the measured properties of molecules with their structural features was made in 1947. In that year, Wiener [121,122] introduced two parameters designed for this purpose. The first of these was termed the path number and was defined as the "sum of the distances between any two carbon atoms in the molecule, in terms of carbon-carbon bonds." A simple algorithm was given for the calculation of this number and it was shown that its value for normal alkanes assumes the form $\frac{1}{6}(n^3 - n)$. The second parameter was called the polarity number and was defined as "the number of pairs of carbon atoms which are separated by three carbon-carbon bonds"; it took the general value $n - 3$ for normal alkanes. Wiener proposed that the variation of any physical property for an isomeric structure as compared to a normal alkane would be given by the linear expression:

$$P(G) = a\Delta w + b\Delta p, \tag{11}$$

where the delta represents the difference in value between that for the normal alkane and its branched counterpart, w is the path number, p the polarity number, and a and b are constants. This relationship was used to correlate several physicochemical properties, including boiling points, heats of vaporization, and surface tensions. Subsequent applications of these parameters have been discussed in the review of Rouvray [123].

The term topological index was first coined in 1971 by Hosoya [124] who used it to describe a new index he introduced that year. This index was defined in terms of the number of 1-coverings of the chemical graph, and is given by the equation:

$$Z(G) = \sum_{k=0}^{[n/2]} \Theta(G,k),\qquad(12)$$

where $\Theta(G,k)$ is the number of ways in which k disconnected K_2 graphs can be imbedded in G as subgraphs, and $[n/2]$ represents the maximal value assumed by the integer k. From this definition, it is evident that $\Theta(G,0)$ will be unity, $\Theta(G,1)$ will represent the number of edges in G, and $\Theta(G,n/2)$ will be the number of 1-factors (also known as Kekulé structures) in G. The index has found many different applications in various physical and chemical settings, including the modeling of physicochemical properties of hydrocarbon species [124], the study of dimer statistics [125], the coding of chemical graphs [126], and the prediction of the π-electron structure of unsaturated hydrocarbon species [127]. For further details on the uses of the index, the reader is referred to the review by Hosoya [128]. Since its inception in 1971, the term topological index has been extended to all graph invariants which exist in the form of scalar numerical parameters.

The concept of differentiating the edges in a chemical graph according to the valences of the two end vertices was first proposed in 1947 by Hartmann [129]. This concept was taken up in 1975 by Randić [130] in his design of a topological index that would be able to deal effectively with the problem of characterizing branched chemical species. Randić started by classifying the edges in a chemical graph into so-called (m,n)-edge types, where m and n represent the degrees of the two vertices in question. Each vertex was then assigned a weighting factor of the form $(1/v)^{\frac{1}{2}}$, where v is the valence of the vertex considered. This weighting factor was selected to give the correct ordering of alkane isomers based on their boiling points. By summing all the weighted edges of the chemical graph, the following index results:

$$\chi(G) = \sum_{\text{edges}} (v_i v_j)^{-\frac{1}{2}}\qquad(13)$$

where v_i and v_j represent respectively the valences of vertices i and j of some given edge in the graph. Because of its great usefulness, this index has since been extended in two important ways, viz. (i) the index has been generalized into a series of indices for which summations are made over subgraphs of G other than the edges [131,132]; and (ii) allowance has been made for heteroatoms in the graph by weighting such atoms according to their number of valence electrons [132].

These indices introduced by Randić [130] have today become known as the set of molecular connectivity indices. The indices have been more widely used than any of the other topological indices put forward to date. The applications of these indices range from correlations of the physical to the biological properties of chemical species. Two monographs by Kier and Hall [132,133] have been devoted solely to discussion of the manifold uses of these indices; a review by Rouvray [134] has focused on the major biological applications. A number of reviews that have treated the applications of topological indices in general have included substantial material on the molecular connectivity indices. Reviews that are very helpful in this regard are those of Balaban *et al* [135], Sabljić and Trinajstić [136], Basak *et al* [137], and Rouvray [138]. Further information can also be obtained from the books of Trinajstić [62] and Bonchev [139]. Although over 100 different topological indices have been proposed up to the present, only a handful of them have so far found widespread use.

1.9 Elementary Bonding Theory

From the earliest times, there has existed an intimate relationship between chemical bonding theory and simple graph-theoretical concepts. As mentioned in Section 2, the first graphs representing individual chemical species drawn by Higgins [20] display the alleged forces operating between pairs of atoms in the species (see Figure 2). Moreover, the first use by Couper [44] of a graphical edge to depict a valence bond (see Figure 8) further attests to the close interaction of the two fields. This interaction has continued down to the present time. The origins of the modern electronic theory of bonding date from the discovery of the electron in 1897 [140,141]; the modern theory is thus exclusively a twentieth century phenomenon. In 1904 Thomson conceived of the chemical bond as a "unit Faraday tube connecting charged atoms in the molecule" [142]. Other important early ideas were those of Kossel [143], who expressed the idea that chemical bonds could vary from the extremes of being either purely ionic or pure covalent; Lewis [144], who explained the difference between ionic and covalent bonds in terms of his cubic model of the atom; and Langmuir [145], who emphasized the role of the stable octet for the electronic shells of atoms in molecules. A comprehensive account of this pioneering work is to be found in the review of Bykov [146].

The advent of quantum theory changed fundamentally our understanding of the nature of the chemical bond. Chemical linkages came to be viewed as a build up of electronic charge density between a pair of atoms, though chemical bonds continued to be represented by graphical edges. The interactions taking place at the electronic level were realized to be far more complex than originally envisaged, and new mathematical tools had to be developed to describe the behavior of the electrons in molecules. Principal among these tools was an approximate equation enunciated by Schrödinger [147] in 1926, which yielded both the energies of electrons in chemical bonds and their probability distributions. It was quickly realized that this equation could not be solved analytically, except in one or two extremely simple cases. This fact led to the development of two major avenues of approach, both of which rested upon approximations: the valence bond and molecular orbital approaches. These two approaches were associated respectively with the names of two American chemists, viz. Pauling and Mulliken. In 1928 Bloch [148] introduced his method of combining orbitals to study delocalized electrons in metals. His linear combination of atomic orbitals was soon to have important consequences in the study of small organic molecules.

Bloch's treatment was incorporated in the molecular orbital study of the benzene molecule by Hückel [149,150]. The latter showed that the energies of the π-electrons in unsaturated molecules such as benzene could be approximated by solving secular determinants containing the Coulomb integral, α, for a carbon atom and the resonance integral, β, for a pair of carbon atoms. In the case of benzene, the determinant assumes the form shown in Figure 16.

Several years later it was shown [151] that this determinant is isomorphic to the adjacency matrix of the molecular graph. The eigenvectors of both matrices will therefore be identical, and the energy levels derived from the Hückel determinant will differ from the eigenvalues of the adjacency matrix only by a scaling factor. What is widely referred to today by the chemist as Hückel theory is thus seen to be equivalent to the spectral theory of the graph theorist. An important conclusion that follows from this observation is that the underlying topology of a molecule determines in a very fundamental way the behavior of that molecule. This realization has led to the generation of a number of theorems of a

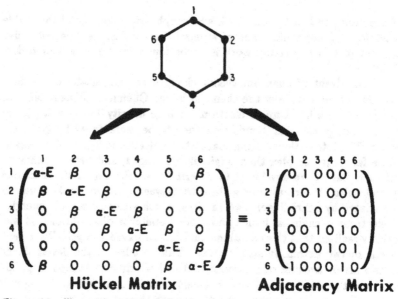

Figure 16. Illustration of the Hückel and adjacency matrices for the benzene molecule showing the closely related structure of each.

graph-theoretical nature which retain their validity even when the Hückel approximations are superseded.

One of the most interesting and important of these theorems is known as the Pairing Theorem. In his study of the polyenes, C_nH_{n+2}, and annulenes, C_nH_n, Hückel demonstrated [149] that their energy levels always exist in the form of plus and minus pairs. This complimentarity implies that whenever such a molecule has an energy level E_k it must also have the energy level $-E_k$. This idea was generalized in 1940 when Coulson and Rushbrooke [152] showed that any "alternant hydrocarbon" would exhibit pairing of its energy levels. In graph-theoretical language this is equivalent to stating that the spectrum of any bipartite graph always exists as paired eigenvalues. The theorem provides a classic example of the interaction of chemistry and graph theory. Whereas the theorem was first adumbrated in a purely mathematical context [153], it was first proved in a chemical paper [152]. Many mathematicians have since rediscovered and developed the theorem in a variety of ways. Details on the early history of this theorem are available in a review by Rouvray [154]; moreover, this review also discusses pioneering work on a number of other fundamental theorems having graph-theoretical content.

The theory of resonance has continued to play a role in quantum chemical thinking and has provided a rich source of inspiration for the application of graph-theoretical techniques. Resonance theory was introduced in a review by Pauling [156] in 1928. In that year the same author also introduced the concept of hybridization of the carbon atom orbitals and ascribed the tetrahedral arrangement of the bonds of carbon atoms to the "resonance phenomenon"[157]. In 1933 Pauling and Wheland [158] began to calculate resonance energies using the valence bond approach. The concept of a canonical collection of structures with noninteracting valences which when superimposed would yield the electronic state of a molecule is due to Rumer [159] and Pauling [160]. A method for determining the number of canonical structures for each degree of excitation, published by Wheland [161] in 1935, led to the introduction of the so-called Wheland polynomial. The uses of polynomials in chemical graph theory are discussed fully by Gutman in chapter 3 of this volume of *Mathematical Chemistry*. For treatments of the modern applications of graph theory in the area of bonding theory, the reader is referred to the reviews by Gutman and Trinajstić [162], Rouvray [154], and Trinajstić, published elsewhere in this volume of *Mathematical Chemistry*.

1.10 Conclusion

The present chapter has sought to propagate the view that the mathematical discipline of graph theory has proved to be and will likely continue to be highly relevant to a wide range of chemical problems and endeavors. Although limitations of space, and the fact that our focus of interest has been a purely historical one, have meant that only certain areas could be discussed here, sufficient has probably been said to establish the credentials of graph theory in the modeling of chemical phenomena. Moreover, it has been pointed out that chemical graph theory in various guises has been around for over two centuries and that in many ways it appears to be ideally suited to application in the chemical context. The earliest application in the eighteenth century involved the depiction of chemical interactions. In the nineteenth century, graphs began to be employed in the solution of a succession of concrete problems, such as the enumeration of chemical isomers. During the twentieth century, the applications of graph theory to chemistry have become so numerous and sophisticated that it is now no longer possible to review all of them comprehensively in anything less than a fullscale treatise. The growth in the popularity of chemical graph theory in recent years is attested to by the annual increase in the numbers of papers being published in the field, as illustrated in Figure 13. The future prospects of graph theory in the chemical domain appear bright and further growth of the field seems assured. Graph theory may well ultimately prove to be one of the most powerful mathematical tools at the disposal of the chemist.

1.11 References

1. E.P. Wigner, Comm. Pure Appl. Math. 13 (1960), 1.
2. R.G. Jahn and B.J. Dunne, Found. Physics 16 (1986), 721.
3. M. Polanyi, *Knowing and Being*, University of Chicago Press, Chicago 1969, p. 219.
4. H. Primas, *Chemistry, Quantum Mechanics and Reductionism*, Springer, Berlin 1983, p. 33.
5. H. Weyl, *Philosophy of Mathematics and Natural Science*, Princeton University Press, Princeton, New Jersey 1949, p. 26.
6. G. Chartrand (ed.), J. Graph Theory (Special Issue) 10(3), (1986).
7. L. Euler, Comment. Acad. Scient. Imper. Petropolitanae 8 (1736), 128.
8. G.R. Kirchhoff, Ann. Phys. Chem. 72 (1847), 497.
9. A. Cayley, Rept. Brit. Assoc. Adv. Sci. 257 (1875).
10. N.L. Biggs, E.K. Lloyd and R.J. Wilson, *Graph Theory: 1736-1936*, Clarendon Press, Oxford 1976.
11. R.J. Wilson, in *Graph Theory with Applications to Algorithms and Computer Science* (ed. Y. Alavi *et al*), Wiley, New York 1985, p. 789.
12. R.J. Wilson, J. Graph Theory 10 (1986), 265.
13. J.J. Sylvester, Nature 17 (1878), 284.
14. F. Cajori (ed.), *Sir Isaac Newton's Principles of Natural Philosophy and his System of the World*, University of California Press, Berkeley, California 1962, p. xviii.
15. P.J. Macquer, *Dictionnaire de Chymie*, Lacombe, Paris 1766.
16. L.L. Whyte, in *Roger Joseph Boscovich* (ed. L.L. Whyte), Allen and Unwin, London 1961, p. 102.
17. H.M. Leicester (ed.), *Mikhail Vasilevich Lomonosov on the Corpuscular Theory*, Harvard University Press, Cambridge, Massachusetts 1970, p. 209.

18. A. Thackray, *Atoms and Powers*, Harvard University Press, Cambridge, Massachusetts 1970, p. 226.
19. M.P. Crosland, Annals Sci. 15 (1959), 75.
20. W. Higgins, *A Comparative View of the Phlogistic and Antiphlogistic Theories*, Murray, London 1789.
21. E.R. Atkinson, J. Chem. Educ. 17 (1940), 3.
22. W.H. Wollaston, Phil. Trans. Roy. Soc. London 98 (1808), 96.
23. W.V. Farrar, in *John Dalton and the Progress of Science* (ed. D.S.L. Cardwell), Manchester University Press, Manchester 1968, Chap. 17.
24. J. Dalton, *New System of Chemical Philosophy*: Vol. 1, Bickerstaff, London 1808, plate 4.
25. N.W. Fisher, Ambix 21 (1974), 29.
26. G.E. Hein, in *Kekulé Centennial*, Advances in Chem. Series 61, Amer. Chem. Soc., Washington, D.C. 1966, chap. 1.
27. H.A.M. Snelders, J. Chem. Educ. 51 (1974), 2.
28. A. Laurent, *Méthode de Chimie*, Mallet-Bachelier, Paris 1854, p. 408.
29. J. Loschmidt, *Chemische Studien*, Gerold, Vienna 1861. Reprinted as Ostwald's Klassiker No. 190, Engelmann, Leipzig 1913.
30. A. Kekulé, Ann. Chem. 137 (1866), 129.
31. C. Graebe, Ann. Chem. 149 (1869), 1.
32. C. Graebe and C. Liebermann, Chem. Ber. 1 (1868), 49.
33. A. Werner, Z. Anorg. Chem. 3 (1893), 267.
34. P. Karrer, Helv. Chem. Acta 3 (1920), 196.
35. H.K. Clark and J.L. Hoard, J. Amer. Chem. Soc. 65 (1943), 2115.
36. W.N. Lipscomb, *Boron Hydrides*, Benjamin, New York 1963.
37. L. Pauling, in *Hydrogen Bonding* (ed. D. Hadzi), Pergamon, London 1958, p. 3.
38. H.W. Kroto, J.R. Heath, S.C. O'Brien, R.F. Curl, and R.E. Smalley, Nature 318 (1985), 162.
39. A.M. Butlerov, Zeitschr. Chem. Pharm. 4 (1861), 549.
40. D.F. Larder, Ambix 19 (1971), 26.
41. A.J. Rocke, Ambix 30 (1983), 1.
42. C.A. Russell, *The History of Valency*, Leicester University Press, Leicester 1971,
43. V.I. Kuznetsov (ed.), *Theory of Valency in Progress*, Mir Publishers, Moscow 1980.
44. A.S. Couper, Ann. Chim. Phys. 53 (1858), 469.
45. A. Crum Brown, *The Theory of Chemical Combination*, M.D. Thesis, University of Edinburgh 1861.
46. A. Crum Brown, Trans. Roy. Soc. Edin. 23 (1864), 707.

47. F.A. Kekulé, Bull. Acad. Roy. Belg. 19 (1865), 551.

48. E. Frankland, *Lecture Notes for Chemical Students*, Van Voorst, London 1866.

49. A. Laurent, Rev. Scient. 14 (1843), 314.

50. C. Gerhardt, Ann. Chim. Phys. [3] 7 (1843), 129.

51. F.A. Kekulé, Ann. Chem. Pharm. 106 (1858), 129.

52. A. Cayley, Phil. Mag. 47 (1874), 444.

53. J.J. Sylvester, Amer. J. Math. 1 (1878), 64.

54. O.J. Lodge, Phil. Mag. 50 (1875), 367.

55. N.S. Kurnakov, Zeitschr. Anorg. Allg. Chem. 169 (1928), 113.

56. M. Frèrejacque, Bull. Soc. Chim. France 6 (1939), 1008.

57. A.T. Balaban, Rev. Roumaine Chim. 18 (1973), 635.

58. D.H. Rouvray, J. Chem. Educ. 52 (1975), 768.

59. D.H. Rouvray, Roy. Inst. Chem. Rev. 4 (1971), 173.

60. D.H. Rouvray and A.T. Balaban, in *Applications of Graph Theory* (ed. L.W. Beineke and R.J. Wilson), Academic Press, London 1979.

61. A.T. Balaban, J. Chem. Inf. Comput. Sci. 25 (1985), 334.

62. N. Trinajstić, *Chemical Graph Theory* (2 vols), Chemical Rubber Co. Press, Boca Raton, Florida 1983.

63. A.T. Balaban (ed.), *Chemical Applications of Graph Theory*, Academic Press, London 1976.

64. R.B. King (ed.), *Chemical Applications of Topology and Graph Theory*, Elsevier, Amsterdam 1983.

65. R.B. King and D.H. Rouvray (ed.), *Graph Theory and Topology in Chemistry*, Elsevier, Amsterdam 1987.

66. H.A.M. Snelders, Chem. Weekblad 60 (1964), 217.

67. J.J. Berzelius, Pogg. Ann. 19 (1830), 326.

68. F.A. Kekulé, *Lehrbuch der Organischen Chemie*, (2 vols), Enke, Erlangen 1861.

69. A.M. Butlerov, Zeitschr. Chem. 5 (1862), 298.

70. L.Pasteur, Comptes Rend. Paris 31 (1850), 480.

71. J.A. Le Bel, Bull. Soc. Chim. France 22 (1874), 337.

72. J.H. van't Hoff, *Voorstel tot Uitbreiding der ... Structuur Formules in de Ruimte*, Greven, Utrecht, Netherlands 1874.

73. D.H. Rouvray, Chem. Soc. Rev. 3 (1974), 355.

74. D.H. Rouvray, Endeavour 34 (1975), 28.

75. A. Crum Brown, J. Chem. Soc. 18 (1865), 230.

76. F.M. Flavitsky, J. Russ. Chem. Soc. 3 (1871), 160.

77. J.V. Knop, W.R. Müller, K. Szymanski, and N. Trinajstić, *Computer Generation of Certain Classes of Molecules*, SKTH Kemija u Industriji, Zagreb, Yugoslavia 1985, p. 41.
78. H.R. Henze and C.M. Blair, J. Amer. Chem. Soc. 53 (1931), 3042.
79. H.R. Henze and C.M. Blair, J. Amer. Chem. Soc. 54 (1932), 1098.
80. H.R. Henze and C.M. Blair, J. Amer. Chem. Soc. 56 (1934), 157.
81. A. Cayley, Phil. Mag. 13 (1857), 172.
82. A. Cayley, Phil. Mag. 3 (1877), 34.
83. F. Hermann, Chem. Ber. 13 (1880), 792.
84. A.C. Lunn and J.K. Senior, J. Phys. Chem. 33 (1929), 1027.
85. J.H. Redfield, Amer. J. Math. 49 (1927), 433.
86. R.A. Davidson, J. Amer. Chem. Soc. 103 (1981), 312.
87. G. Pólya, Acta Math. 68 (1937), 145.
88. A.T. Balaban, J. Labelled Compds 6 (1970), 211.
89. A.T. Balaban, Croat. Chem. Acta 51 (1978), 35.
90. A.T. Balaban, Rev. Roumaine Chim. 20 (1975), 227.
91. D.H. Rouvray, J. S. Afr. Chem. Inst. 26 (1973), 141.
92. J.E. Leonard, *Studies in Isomerism: Permutations, Point Group Symmetries, and Isomer Counting*, Ph.D. Thesis, California Inst. Technol., California 1971.
93. N.E. De Bruijn, J. Comb. Theory 2 (1967), 418.
94. E. Ruch, W. Hässelbarth, and B. Richter, Theor. Chim. Acta 19 (1970), 288.
95. K. Balasubramanian, Chem. Rev. 85 (1985), 599.
96. C. Gerhardt, Ann. Chim. 14 (1845), 107.
97. H. Kopp, Ann. Phys. 47 (1839), 133.
98. H. Kopp, Ann. Chem. 41 (1842), 79.
99. H. Kopp, Ann. Chem. 41 (1842), 169.
100. H.S. Redgrove, Chem. News (London) 116 (1917), 37.
101. K. Fajans, Chem. Ber. 53 (1920), 643.
102. C.T. Zahn, J. Chem. Phys. 2 (1934), 671.
103. T.L. Allen, J. Chem. Phys. 31 (1959), 1039.
104. G.R. Somayajulu, A.P. Kudchadker, and B.J. Zwolinski, Ann. Rev. Phys. Chem. 16 (1965), 213.
105. E.A. Smolenski, Russ. J. Phys. Chem. 38 (1964), 700.
106. M. Gordon and J.W. Kennedy, J. Chem. Soc. Faraday Trans. II 69 (1973), 484.
107. G. Calingaert and J.W. Hladky, J. Amer. Chem. Soc. 58 (1936), 153.
108. S.S. Kurtz and M.R. Lipkin, Ind. Eng. Chem. 33 (1941), 779.

109. H. Kopp, Ann. Chem. Pharm. 50 (1844), 71.
110. D.H. Rouvray, Chem. Tech. 3 (1973), 378.
111. S. Sugden, J. Chem. Soc. (1924), 1177.
112. R. Samuel, J. Chem. Phys. 12 (1944), 167.
113. B. Nekrassow, Zeitschr. Phys. Chem. A140 (1929), 342.
114. B. Nekrassow, Zeitschr. Phys. Chem. A141 (1929), 378.
115. W.J. Taylor, J.M. Pignocco, and F.D. Rossini, J. Res. Nat. Bur. Stand. 34 (1945), 413.
116. S.A. Mumford, J. Chem. Soc. (1952), 4897.
117. D.H. Rouvray and R. Pandey, J. Chem. Phys. 85 (1986), 2286.
118. K.J. Laidler, Can. J. Chem. 34 (1956), 626.
119. V.M. Tatevski and Y.G. Papulov, Russ. J. Phys. Chem. 34 (1960), 115.
120. G.R. Somayajulu and B.J. Zwolinski, Trans Far. Soc. 62 (1966), 2327.
121. H. Wiener, J. Amer. Chem. Soc. 69 (1947), 17.
122. H. Wiener, J. Phys. Chem. 52 (1948), 1082.
123. D.H. Rouvray, in *Mathematics and Computational Concepts in Chemistry*, (ed. N. Trinajstić), Horwood Publishers, Chichester, U.K. 1986, p. 295.
124. H. Hosoya, Bull. Chem. Soc. Japan 44 (1971), 2332.
125. H. Hosoya and A. Motoyama, J. Math. Phys. 26 (1985), 157.
126. H. Hosoya, J. Chem. Docum. 12 (1972), 181.
127. H. Hosoya, K. Hosoi, and I. Gutman, Theor. Chim. Acta 38 (1975), 37.
128. H. Hosoya, in *Mathematics and Computational Concepts in Chemistry*, (ed. N. Trinajstić), Horwood Publishers, Chichester, U.K. 1986, p. 110.
129. H. Hartmann, Zeitschr. Naturforsch. 2A (1947), 259.
130. M. Randić, J. Amer. Chem. Soc. 97 (1975), 6609.
131. L.B. Kier, W.J. Murray, M. Randić, and L.H. Hall, J. Pharm. Sci. 65 (1976), 1226.
132. L.B. Kier and L.H. Hall, *Molecular Connectivity in Chemistry and Drug Research*, Academic Press, New York 1976.
133. L.B. Kier and L.H. Hall, *Molecular Connectivity in Structure-Activity Analysis*, Research Studies Press, Letchworth, U.K. 1986.
134. D.H. Rouvray, Acta Pharm. Jugosl. 36 (1986), 239.
135. A.T. Balaban, A. Chiriac, I. Motoc, and Z. Simon, Lect. Notes in Chem. 15 (1980), 22.
136. A. Sabljić and N. Trinajstić, Acta Pharm. Jugosl. 31 (1981), 189.

137. S.C. Basak, V.R. Magnuson, G.J. Niemi, R.R. Regal, and G.D. Veith, Proc. Fifth Int. Conf. Math. Modeling (eds. X.J.R. Avula, G. Leitman, C.D. Mote, and E.Y. Rodin), Pergamon Press, New York 1986.

138. D.H. Rouvray, J. Comput. Chem. 8 (1987), 470.

139. D. Bonchev, *Information Theoretic Indices for Characterization of Chemical Structures*, Research Studies Press, Chichester, U.K. 1983.

140. E. Wiechert, Schriften Phys.-Ökon. Gesell. Königsberg 38 (1897), 3.

141. J.J. Thomson, Phil. Mag. [5] 44 (1897), 293.

142. J.J. Thomson, Phil. Mag. [6] 7 (1904), 237.

143. W. Kossel, Ann. Phys. 49 (1916), 229.

144. G.N. Lewis, J. Amer. Chem. Soc. 38 (1916), 762.

145. I. Langmuir, J. Amer. Chem. Soc. 41 (1919), 868.

146. G.V. Bykov, Chymia 10 (1965), 199.

147. E. Schrödinger, Ann. Phys. 79 (1926), 360.

148. F. Bloch, Zeitschr. Phys. 52 (1928), 555.

149. E. Hückel, Zeitschr. Phys. 70 (1931), 204.

150. E. Hückel, Zeitschr. Phys. 76 (1932), 628.

151. K. Ruedenberg, J. Chem. Phys. 22 (1954), 1878.

152. C.A. Coulson and G.S. Rushbrooke, Proc. Cambridge Phil. Soc. 36 (1940), 193.

153. O. Perron, Math. Ann. 64 (1907), 248.

154. D.H. Rouvray, in *Chemical Applications of Graph Theory*, (ed. A.T. Balaban), Academic Press, London 1976, p. 175.

155. H.H. Günthard and H. Primas, Helv. Chim. Acta 39 (1956), 1645.

156. L. Pauling, Chem. Rev. 5 (1928), 173.

157. L. Pauling, Proc. Natl. Acad. Sci. 14 (1928), 359.

158. L. Pauling and G.W. Wheland, J. Chem. Phys. 1 (1933), 362.

159. G. Rumer, Nachr. Ges. Wiss. Göttingen, Math. Phys. Kl.337 (1932).

160. L. Pauling, J. Chem. Phys. 1 (1933), 280.

161. G.W. Wheland, J. Chem. Phys. 3 (1935), 356.

162. I. Gutman and N. Trinajstić, Topics Curr. Chem., 42 (1973), 49.

Chapter 2

ELEMENTS OF GRAPH THEORY FOR CHEMISTS

Oskar E. Polansky

Max-Planck-Institut für Strahlenchemie, D-4330 Mülheim a. d. Ruhr,
Federal Republic of Germany

Translated by Vinzenz Bachler

Max-Planck-Institut für Strahlenchemie, D-4330 Mülheim a. d. Ruhr,
Federal Republic of Germany

In recent decades the literature on graph theory has increased by leaps and bounds. Here, only a minor selection of the extensive material available will be presented. Useful supplements are the textbooks and monographs listed at the end of this chapter.

2.1   What is a Graph and what kinds of Graph exist?

Any graphical representation composed of points and lines is called a graph. Airline networks and other traffic connections are common examples of graphs. Additionally, chemists are familiar with graphs corresponding to the constitutional formulas of chemical compounds. Thus, the graph depicted below represents not only the constitutional formula of iso-pentane but also those of dimethyl-ethylchlorosilane and other compounds; these are said to be iso-topological (see Section 12).

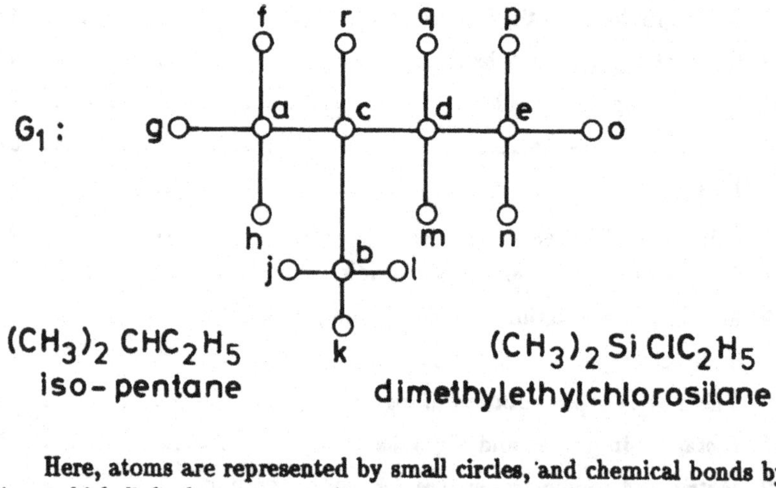

G_1 :

$(CH_3)_2 CHC_2H_5$
iso-pentane

$(CH_3)_2 Si ClC_2H_5$
dimethylethylchlorosilane

Here, atoms are represented by small circles, and chemical bonds by lines which link the corresponding circles. In graph theory the objects designated by small circles are called *vertices, points* or *nodes*; the use of the latter term is not to be recommended, however, because in quantum chemistry, surfaces with wave functions having zero values are called nodes. The linking lines are denoted as *edges, lines,* or when directed, as *arcs.* Thus, the structures defined in mathematics as *graphs consist of two sets: a vertex set and an edge set.* The elements of the edge set represent relations between vertex pairs. This statement can be expressed as follows: *a graph consists of a vertex set on which a pair relation $\mathcal{E}$ is defined.*

$$\mathcal{G} = [\mathcal{V}, \mathcal{E}]. \tag{1}$$

The vertex set $\mathcal{V}$ can be empty, of finite or infinite cardinality. The corresponding graphs are respectively referred to as empty, finite or infinite.

By means of the pair relationship a set of vertex pairs, e.g. $(b, c) \in \mathcal{G}_1$, is defined. These vertex pairs can be ordered or unordered. Unordered vertex pairs are called edges; in the following they are denoted as $(b, c) \in \mathcal{G}_1$, and are represented by a line connecting the corresponding vertices. The geometric form of such a line is meaningless. Ordered vertex pairs, however, are called arcs and will be designated by square brackets and denoted by a directed line, where the arrow points from the first to the second vertex of the pair, e.g. $\langle u, v \rangle \in \mathcal{G}_2$. Here, the vertex u is the starting point (*source*) and vertex v the final point (*sink*) of the arc $\langle u, v \rangle$. Moreover, the set $\mathcal{E}$ may include elements which relate a vertex to itself, e.g. $(z, z) \in \mathcal{G}_5$; such elements are called *loops*.

Thus for a graph having more than two vertices, one of the following relationships between vertex pairs will hold:

(i) The two vertices are unpaired, meaning that in the graphical representation of $\mathcal{G}$ neither an edge nor an arc is found between these two vertices;

(ii) They form an unordered pair, i.e. an edge;

(iii) They form an ordered pair, i.e. an arc;

(iv) The vertices occur as unordered pairs more than once, i.e. they are the endpoints of a multiple edge;

(v) The vertices occur as ordered pairs more than once and are directed (in the same or opposite sense), i.e. they are the endpoints of a multiple arc.

Any edge (s, t) may be thought of as a superposition of two oppositely directed arcs $\langle s, t \rangle$ and $\langle t, s \rangle$. Consequently, we may exclude in general the simultaneous occurrence of ordered and unordered vertex pairs in one graph. This exclusion does not affect the variety of types of graphs. The family of representative graphs of polymers (see Section 14) represents an exception to this rule.

Graphs comprised only of arcs, such as $\mathcal{G}_2$ and $\mathcal{G}_3$, are referred to as *directed graphs* or *digraphs*. If any vertex pair is associated with only one arc, as in G_3, the graph is called an *oriented graph*.

For graphs having no arcs, we distinguish three types: *Simple graphs* (often called *graphs* for short) contain no multiple edges or loops, like G_1. In *multigraphs*, multiple edges but no loops are permitted, cf. G_4. Graphs which contain loops are called *pseudographs*; they may, however, also contain multiple edges like G_5.

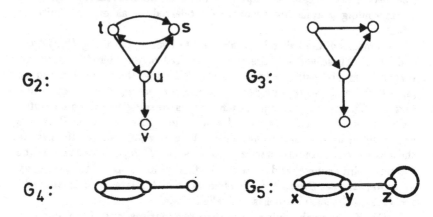

In the following sections we discuss in the main the properties of simple graphs. Generally, the results derived hold also for the other types of graphs; occasional peculiarities and violations of this general rule are pointed out. Of course, when using digraphs, the term "edge" must be replaced by "arc".

It is often convenient to label the vertices and/or the edges of a graph by letters or numbers. Such graphs are called *labelled graphs*; G_1, G_2 and G_5 represent examples of such graphs. The vertex and edge sets of these graphs can be explicitly stated in terms of letters as shown here for G_1:

$$\mathcal{V}_1 = \{a,b,c,d,e,f,g,h,j,k,l,m,n,o,p,q,r\};$$

$$\mathcal{E}_1 = \{(a,c),(a,f),(a,g),(a,h),(b,c),(b,j),(b,k),(b,l),(c,d),(c,r),$$
$$(d,e),(d,m),(d,q),(e,n),(e,o),(e,p)\}.$$

Two graphs G' and G'' are called *isomorphic* if a bijective mapping of the vertex sets exists, i.e. $\mathcal{V}(G') \longleftrightarrow \mathcal{V}(G'')$, which retains the pair relationship $\mathcal{E}(G') \longleftrightarrow \mathcal{E}(G'')$. The bijective mapping of G' onto G'' and vice versa is denoted by $G' \longleftrightarrow G''$, whereas isomorphism of the two graphs is designated by $\mathcal{G}' \cong \mathcal{G}''$. One example of a pair of isomorphic graphs is given below:

The respective numbers of vertices and edges of a graph, i.e. the cardinality of its vertex and edge set, will be designated hereafter by:

$$n = |\mathcal{V}| = n(\mathcal{G}), \qquad m = |\mathcal{E}| = m(\mathcal{G}). \tag{2}$$

The role played by graph theory in chemistry is the subject of this monograph. This can be exemplified by the graph $\mathcal{G}_1$. Graphs corresponding to a constitutional formula are called constitutional graphs; graphs where representation of the hydrogen atoms is suppressed, are termed hydrogen-suppressed or *skeleton graphs*.

2.2 Some Graph-theoretical Terms

For describing details of graphs, an extensive terminology has been developed, comprising (in any language) many synonyms. In this section some of the most important terms are explained and interrelated wherever this is possible.

Two vertices are *adjacent* when they are connected by an edge. The edge and these two vertices are said to be *incident* to one another. Two edges are said to be adjacent provided they have a vertex in common. Single edges of a multiple edge are always adjacent.

A *half edge* is an edge together with one of its incident vertices.

A *wreath of edges* around a given vertex is formed by the vertex in question and all its incident edges.

The number of edges incident with a given vertex is called the degree g of that vertex; for example, one finds in $\mathcal{G}_1$ that $g_a = 4$ and $g_f = 1$, etc. In digraphs two degrees are assigned to any vertex. The *indegree* g^- counts the arcs ending on the vertex; the *outdegree* g^+ counts the arcs originating from the vertex.

Since any edge comprises two vertices, the following condition holds for simple graphs:

$$\sum_{a \in \mathcal{V}} g_a = 2m \tag{3}$$

The number of edges is necessarily a natural number. Consequently, the number of vertices with odd degree must be even. In chemistry, this condition is known as "the paired number law for atoms".

Because each arc increases the out- and indegree by 1 for digraphs,

$$\sum_{a \in \mathcal{V}} g_a^+ = \sum_{a \in \mathcal{V}} g_a^- \tag{4}$$

Vertices of degree 1 are called *terminal vertices*; the vertices $\{f, g, \ldots, r\} \in G_1$ are terminal vertices. Edges incident with terminal vertices are called *terminal edges*.

A graph, in which all vertices are of the same degree, g, is called a *regular graph of degree g*.

A *walk*, v_{rs}, is a sequence of pairwise adjacent edges, leading from vertex r to vertex s. Here, any vertex or edge may be traversed several times. The arcs of digraphs, however, may be traversed only along their direction.

A *trail*, z_{rs}, is a walk, v_{rs}, in which a given edge may only be traversed only once.

A *path*, w_{rs}, is a walk, v_{rs}, in which each of its vertices occurs only once. If in $\mathcal{G}$ multiple edges are absent, the path $w_{rs} \in \mathcal{G}$ is uniquely defined by the corresponding sequence of vertices. Thus, $w_{fk} \in \mathcal{G}_1$ and $w_{su} \in \mathcal{G}_2$ are denoted as follows:

$$w_{fk} = (f, a, c, b, k); \quad w_{su} = \langle s, t, u \rangle.$$

If multiple edges occur, such as $(x, y)_i \in \mathcal{G}_5$, $i = 1, 2, 3$, the particular edge of a multiple edge pertaining to the path must be indicated. Thus, because of the multiple edge $(x, y)^3 \in \mathcal{G}_5$, three different paths $w_{xs} \in \mathcal{G}_5$ exist. Since a vertex of a path is permitted to be traversed only once, a loop cannot form part of a path.

A walk closed in itself is called a *self-returning walk*. A self-returning path is termed a *cycle*.

In digraphs, the concepts of *semipath* and *semicycle* are important. Both are closed sequences of pairwise adjacent arcs where their direction is insignificant. A semipath leads from s via u to v and vice versa. Obviously, any directed path (cycle) corresponds to a semipath (semicycle) with the same sequence of vertices. The reverse statement is not necessarily true, as already illustrated by the example given above $(s, u, v) \in \mathcal{G}_2$: no directed path corresponds with this semipath in $\mathcal{G}_2$.

A cycle containing all vertices of a graph is referred to as a *Hamilton cycle.*

An *Eulerian trail* is a trail containing all vertices of a graph; such a trail can be open or closed when G contains two or no vertices of odd degree, respectively. For graphs with more than two vertices of odd degree no Eulerian trail exists.

The *length of a path (cycle)* is the number of edges associated with this path (cycle). The relationship of this concept to the discrete metric which can be defined upon simple graphs is outlined in Section 11.

A graph is called *bipartite* if adjacent vertices always have different colours when two different colours are used for colouring the vertices. It can be shown that bipartite graphs do not contain cycles of odd length; this requirement is a necessary and sufficient condition for bipartite graphs.

2.3 Connectedness of Graphs

A graph $\mathcal{G}$ is called *connected* if for each pair of vertices $\{r, t\} \in \mathcal{G}$ at least one path w_{rt} exists; otherwise $\mathcal{G}$ is a *disconnected graph.*

Any connected part $\mathcal{G}_j$ of a disconnected graph is called a *component* of $\mathcal{G}$ and, obviously, $\mathcal{G} = \bigcup_{\{j\}} \mathcal{G}_j$. Therefore, a connected graph can be designated as a single component graph.

While undirected graphs are either connected or disconnected, in digraphs this concept becomes more subtle: a digraph $\mathcal{D}$ is *strongly connected* or *strong* if for any vertex pair $\{r, t\} \in \mathcal{D}$ at least two oppositely directed paths $\langle r, \ldots, t \rangle$ and $\langle t, \ldots, r \rangle$ exist. The oriented cycle $\mathcal{G}_6$ is an example of a strong digraph. Note that in this example the directed paths $\langle r, \ldots, t \rangle$ and $\langle t, \ldots, r \rangle$ have only their terminal vertices, r and t, in common.

A digraph is *unilaterally connected,* or *unilateral,* if for any pair of vertices at least one vertex can be reached from the other one by a directed path; for instance, graph $\mathcal{G}_2$ is an unilateral digraph.

If two edges in a digraph are connected by a semipath, the digraph is called *weakly connected* or *weak*; graphs $\mathcal{G}_3, \mathcal{G}_7$ and $\mathcal{G}_8$ are weak digraphs.

A digraph which is not even weak is described as *unconnected.*

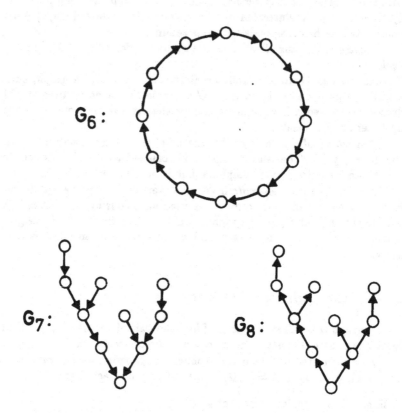

A weak digraph is said to be an *in-tree (out-tree)* if and only if exactly one vertex has outdegree (indegree) 0, and all others have outdegree (indegree) 1. G_7 is an in-tree, and G_8 an out-tree.

The concepts of connectedness for digraphs form a hierarchy, since any strong graph is also unilateral and any unilateral digraph is also weak; the converse statements are not true.

On what does the connectedness of a graph depend? Obviously, a sufficient number of edges, m, must be present. As can be easily proved, to connect n vertices at least $n-1$ edges are needed. Thus, $m \geq n-1$ is a necessary condition; but is it also sufficient? From $n-1$ vertices, where $n > 2$, up to $(n-1)(n-2)/2$ different unordered pairs may be formed, each one representing an edge. Thus, even $(n-1)(n-2)/2$ edges may be arranged in such a manner that only $n-1$ vertices are connected and one of the n vertices is left over. This shows that $n-1 \leq m \leq (n-1)(n-2)/2$ is a necessary but not sufficient condition for the connectedness of a graph. As it also shows, from comparison of the relative magnitudes of the numbers m and n, definite conclusions concerning the connectedness of a graph can be drawn only if either $m < n-1$ or $m > (n-1)(n-2)/2$ holds. For intermediate values of m, $n-1 \leq m \leq (n-1)(n-2)/2$, another concept to establish the connectedness of a graph is needed; such a concept is that of the spanning tree of a graph outlined below.

Let $\mathcal{G}$ be a connected graph with n vertices and $m = n - 1$ edges. $\mathcal{G}$ contains no cycle because the necessary edge is not available for the closure of an open path $w_{rs} \in \mathcal{G}$ into a cycle.

Connected graphs without cycles are called *trees*, and usually denoted by the symbol $\mathcal{T}$. The number of vertices n and edges m of a tree are related by the condition

$$m = n - 1. \tag{5}$$

Any vertex of degree $g > 2$ indicates a branching of the tree. A tree without branching is called a *path graph* $\mathcal{P}_n$; it possesses exactly 2 terminal vertices of degree 1 and $n - 2$ vertices of degree 2. A maximally branched tree is called a *star graph*, $\mathcal{K}_{1,n-1}$, and has $n - 1$ terminal vertices and one vertex of degree $n-1$. Graphs $\mathcal{G}_9$, $\mathcal{G}_{10}$ and $\mathcal{G}_{11}$ all represent nonisomorphic trees having five vertices; they represent the carbon skeletons of n-, iso- and neo-pentane. $\mathcal{G}_9$ and $\mathcal{G}_{11}$ are described as the path $\mathcal{P}_5$ and the star $\mathcal{K}_{1,4}$, respectively.

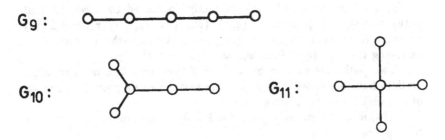

A graph $G = \bigcup_{\{j\}} \mathcal{G}_j$ consisting of $k > 1$ components but without any cycles is called a *forest*. Each of its components is a tree for which eq.(5) holds. By adding these k equations, the condition $m = n - k$ for the forest $\mathcal{G}$ is derived.

Evidently, a graph consisting of k components, n vertices and $m > n - k$ edges contains

$$\mu = m - n + k \qquad (6)$$

independent cycles; μ is called the *cyclomatic number* of G. In order to explain the terms *dependent* and *independent* as applied to cycles, consider graph G_{12} which contains a cycle $w_{rr} \in G_{12}$ and a branched side chain. This chain originates from the cycle w_{rr} at the vertex q adjacent to vertex r. Introducing a new edge into G_{12} by adding either the edge (s,t) or (r,s), produces two new graphs, namely

$$G_{13} = G_{12} \cup (s,t), \qquad G_{14} = G_{12} \cup (r,s).$$

Clearly, by adding the new edge, a new cycle is formed. However, do the graphs G_{13} and G_{14} contain the same number of cycles, namely two? Inspection of G_{13} indicates the presence of only two cycles. Besides the cycle w_{rr} originally present, we observe the cycle w_{tt}. Both cycles have no vertex, and consequently no edge, in common. Therefore, the intersection of these two cycles is empty.

$$w_{rr} \cap w_{tt} = \emptyset \qquad (7)$$

At a first glance we also identify in G_{14} two cycles, i.e. w_{rr} which was originally present, and w_{ss} which is newly formed. The intersection, however, is the edge (q,r):

$$w_{rr} \cap w_{ss} = (q,r) \qquad (8)$$

In G_{14} the vertices q and r are evidently connected by three different paths, namely $(w_{qr})_1 = w_{rr} - (q,r)$, $(w_{qr})_2 = w_{ss} - (q,r)$, and finally the edge (q,r) itself, $(w_{qr})_3 = (q,r)$; the first two paths are formed by removing the edge (q,r) from w_{rr} and w_{ss}, respectively.

Two paths are said to be *disjoint* if they have the same terminal vertices but only those in common. As may be easily proved, the union of two disjoint paths forms a cycle.

Obviously, the paths $(w_{qr})_j$, $j = 1, 2, 3$, are disjoint and, hence, they form pairwise cycles:

$$(w_{qr})_1 \cup (w_{qr})_2 = C_3,$$

$$(w_{qr})_1 \cup (w_{qr})_3 = C_1 = w_{rr},$$

$$(w_{qr})_2 \cup (w_{qr})_3 = C_2 = w_{ss}.$$

From these three cycles only two may be chosen as independent because they satisfy the following relation:

$$C_3 = (C_1 \cup C_2) \setminus (C_1 \cap C_2) \qquad (9)$$

This finding may be generalized as follows: *From any pair of independent cycles, C_1 and C_2, which have a nonempty intersection, $C_1 \cap C_2 \neq \emptyset$, a dependent cycle C_3, is derived according to eq.(9).*

Since in $\mathcal{G}_{13}$ the intersection of the cycles w_{rr} and w_{tt} is empty (see eq.(7)) no dependent cycle can be derived for these cycles.

Without further comment, we now mention two less elementary relations. (1) The independent cycles of $\mathcal{G}$ form a vector space called the cycle basis of $\mathcal{G}$. (2) If between the vertices $\{r, s\} \in \mathcal{G}$ ω disjoint paths are present, then the vertices r and s belong to $\binom{\omega}{2}$ cycles of which $\omega - 1$ cycles are independent.

The number of independent cycles according to eq.(6) is a function of n, m and k. By contrast, the number of dependent cycles is related to the mutual location of the independent cycles in the graph and must be evaluated on its own.

A simple graph on n vertices having the maximal number $m = \binom{n}{2}$ of edges is called a *complete graph* K_n:

$$K_n = [\mathcal{V}(n), \, \mathcal{E}_o(n)] \qquad (10)$$

In such a graph all vertices and edges are mutually adjacent. Consequently, a complete graph is a regular graph of degree $n - 1$. Moreover, K_n contains the maximal number of cycles. The graphs $\mathcal{G}_{15}$ to $\mathcal{G}_{20}$ are the complete graphs K_n, where $n = 1, 2, \ldots, 6$.

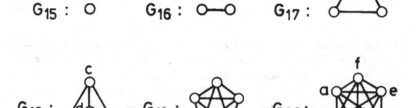

The completeness of $\mathcal{K}_n$ is a consequence of the completeness of $\mathcal{E}_o(n)$, and comprises all the unordered pairs which can be formed from the n vertices of $\mathcal{V}(n)$. Consequently, any graph with n vertices,

$$\mathcal{G} = [\mathcal{V}(n), \mathcal{E}(\mathcal{G})], \tag{11}$$

can be produced from $\mathcal{K}_n$ by removing the appropriate edges. Thus,

$$\mathcal{G} \subset \mathcal{K}_n \tag{12}$$

and $\mathcal{K}_n$ *is the supergraph of all graphs consisting of n vertices.*

If all the removed edges are collected in the set $\mathcal{E}(\overline{\mathcal{G}})$, then obviously the following relations will hold:

$$\mathcal{E}_o(n) = \mathcal{E}(\mathcal{G}) \cup \mathcal{E}(\overline{\mathcal{G}}), \tag{13}$$

$$\mathcal{E}(\mathcal{G}) \cap \mathcal{E}(\overline{\mathcal{G}}) = \emptyset. \tag{14}$$

According to eq.(1) and using the sets $\mathcal{V}(n)$ and $\mathcal{E}(\overline{\mathcal{G}})$, a further graph may be defined:

$$\overline{\mathcal{G}} = [\mathcal{V}(n), \mathcal{E}(\overline{\mathcal{G}})], \tag{15}$$

which is called the *complement* of $\mathcal{G}$. In $\mathcal{K}_4 = \mathcal{G}_{18}$ several $\mathcal{P}_4$ graphs are contained, whose complements are also $\mathcal{P}_4$. Graphs having this property are called self-complementary.

An important subgraph of $\mathcal{K}_n$ is the *cycle* C_n, this being a regular graph of degree two.

A complete *bipartite graph* $\mathcal{K}_{n',n''}$ comprises n' vertices of one colour and n'' vertices of another colour and all the edges which connect vertices

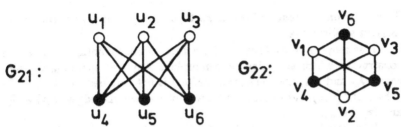

of different colours. The above mentioned star $K_{1,n-1}$ is a bipartite graph. An important complete bipartite graph is $K_{3,3} = \mathcal{G}_{21} = \mathcal{G}_{22}$; the isomorphism $\mathcal{G}_{21} \cong \mathcal{G}_{22}$ is a consequence of $u_j \longleftrightarrow v_j$ being valid for all vertices.

If for the vertex and edge sets of two graphs $\mathcal{G}$ and $\mathcal{G}'$ the relations

$$V(\mathcal{G}') \subseteq V(\mathcal{G}) \quad \text{and} \quad \mathcal{E}(\mathcal{G}') \subseteq \mathcal{E}(\mathcal{G})$$

are valid, then $\mathcal{G}'$ is called a *subgraph* of $\mathcal{G}$, and $\mathcal{G}$ the *supergraph* of $\mathcal{G}'$. Provided $V(\mathcal{G}') = V(\mathcal{G})$ holds, $\mathcal{G}'$ is said to be a *spanning subgraph* of $\mathcal{G}$.

In the discussion of connectedness, the trees spanning a graph $\mathcal{G}$ are of major importance. Such *spanning trees* are derived from $\mathcal{G}$ by removing definite edges; their number is given by the cyclomatic number, μ, eq.(6). As an example, $\mathcal{G}_{23}$ is a spanning tree of $\mathcal{G}_{12}$, $\mathcal{G}_{13}$ and $\mathcal{G}_{14}$.

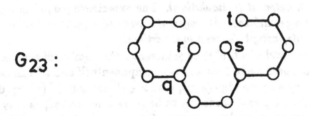

The vertex and edge sets of a connected graph, $\mathcal{G} = [V(\mathcal{G}), \mathcal{E}(\mathcal{G})]$, and one of its spanning trees, $T = [V(T), \mathcal{E}(T)]$, are interrelated by $V(T) = V(\mathcal{G})$ and $\mathcal{E}(T) \subseteq \mathcal{E}(\mathcal{G})$. The edge sets are identical only when $\mathcal{G}$ is a tree itself, i.e. it is the only tree spanning $\mathcal{G}$.

If $\mathcal{G}$ is disconnected, the spanning subgraphs are also disconnected and must consist of at least the same number of components as $\mathcal{G}$. Suppose each component of the disconnected graph $\mathcal{G}$ is spanned by a tree. The spanning subgraph of $\mathcal{G}$ is then a forest consisting of the same number of components as $\mathcal{G}$. In turn one may conclude: *A graph is connected if and only if it comprises at least one tree as a spanning subgraph.*

This criterion of connectedness is particularly appropriate for computer-assisted applications.

It is easy to verify that for G_{12} a total of six spanning trees can be constructed which are pairwise isomorphic. Any tree spanning G_{12} also spans G_{13} and G_{14}; the converse statement is not true because $\mathcal{E}(G_{12})$ is a subset of $\mathcal{E}(G_{13})$ and $\mathcal{E}(G_{14})$, and does not contain the edges $(s,t) \in G_{13}$ and $(r,s) \in G_{14}$.

The constitutional graphs of common compounds and those of Möbius compounds and knots (recently synthesized) are connected. The graphs of catenanes and rotaxanes, however, consist of two or more components reflecting the dimeric (oligomeric) character of these compounds.

2.4 Partitioning of a Graph

The removal of edges or vertices (together with their incident edges) from a connected graph leads to a partitioning of the graph into several components, provided that proper edges and vertices are selected. Thus, for example, G_{23} is partitioned into components by removing an arbitrary edge or vertex of degree $g \geq 2$. An edge or a vertex whose removal partitions the connected graph into components is called, respectively, a *bridge* or a *cutpoint (articulation)*. The experience gained in examining G_{23} may be generalized: *All edges of a tree are bridges and all its vertices are cutpoints, except the terminal vertices.*

The removal of a bridge edge increases the number of components by exactly one, because in turn any two components G' and G'' can be always connected by a single edge (x,y), with $x \in G'$ and $y \in G''$ (cf. eq.(6)). The situation for cutpoints turns out to be more complicated, as may be seen in the following example: Removal of an unlabelled vertex of degree two adjacent to $q \in G_{23}$ leads to two components, whilst removal of q itself yields three components.

The vertices incident with a bridge are always cutpoints provided they are not terminal vertices. This statement is an immediate consequence of that given above: If a vertex incident with a bridge is removed, the bridge itself is removed. This is due to the fact that the bridge belongs to the wreath of edges of its incident vertices. However, the results of removing a bridge or one of its incident vertices are different. In the first case, the number of vertices n remains unchanged whereas in the second it is decreased by one.

A graph without cutpoints is called a *block*. Of course, no bridges can occur in a block because their incident vertices would be cutpoints and thus in contradiction to the definition of a block.

Removal of all the bridges in a connected graph is often insufficient to partition it into a spanning subgraph consisting only of blocks because in $\mathcal{G}$ cutpoints not incident with bridges may be present. An example is the vertex $a \in \mathcal{G}_{24}$; $\mathcal{G}_{24}$ is obviously not a block.

$$\mathcal{G}_{24}:$$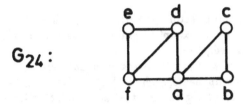

The definition of a bridge implies that removal of an edge which is not a bridge does not eliminate the connectedness of a graph. In $\mathcal{G}_{12}$ all edges contained in the paths w_{qs} and w_{qt} are bridges whereas all edges of the cycle w_{rr} are not bridges. Thus, removal of the edge $(q,r) \in w_{rr}$ does not lead to a partitioning of $\mathcal{G}_{12}$ into components; on the contrary $\mathcal{G}_{12} - (q,r)$ is connected. However, if the edge (q,r) is removed together with another edge of the cycle w_{rr}, $\mathcal{G}_{12}$ is partitioned into two components. The cardinalities of the vertex sets of the components will depend on the choice of the second edge. A set consisting of a minimal number of edges, whose removal increases the number of components by one, is called an *edge cut set*. The two edges of $w_{rr} \in \mathcal{G}_{12}$ form such an edge cut set. They also decompose $\mathcal{G}_{13}$, but not $\mathcal{G}_{14}$. In $\mathcal{G}_{14}$ an edge cut set which contains the edge (q,r) must be of cardinality three.

Similarly, a set with a minimal number of vertices is called a *vertex cut set*, provided its removal increases the number of components by one. Examples are $\{q,r\} \in \mathcal{G}_{14}$ and $\{d,f\} \in \mathcal{G}_{24}$.

Obviously, a bridge or a cutpoint is respectively an edge or a vertex cut set with cardinality one.

Vertex and edge cut sets are respectively subsets of $\mathcal{V}(\mathcal{G})$ and $\mathcal{E}(\mathcal{G})$. Owing to the requirement of minimal cardinality, these subsets do not contain cutpoints or bridges provided their cardinality exceeds one.

The edge cut sets are of major importance for certain expansions of the characteristic and acyclic polynomials of graphs (see Section 9) as well as for the discussion of some topological properties (see Section 12).

In chemistry they are used intuitively in problems, when larger molecules are thought to be composed of smaller entities.

A completely different concept of decomposition is used as the basis for the *factorization* of a connected graph. Factorization is defined as a partitioning of $\mathcal{G}$ into edge-disjoint spanning subgraphs which have definite properties; such a subgraph is called a *factor* of $\mathcal{G}$. The property usually required is the regularity of the subgraph, i.e. that all its vertices have the same degree g. Regular factors of degree g are briefly designated as g-factors.

Obviously, the 0-factor, $\mathcal{G}(n,\emptyset)$, of a graph with n vertices is the complement $\overline{\mathcal{K}}_n$ of the complete graph $\mathcal{K}_n$; $\mathcal{G}(n,\emptyset) = \overline{\mathcal{K}}_n$ is completely unconnected and edgeless.

A 1-factor of a graph $\mathcal{G}$ having $n = 2n'$ vertices is again a graph consisting of n' components, each of which is $\mathcal{K}_2 = \mathcal{G}_{16}$. Because $\mathcal{K}_2$ is the only regular graph of degree one, it is easily seen that for a graph defined by an odd number of vertices, a 1-factor cannot exist. Also, it is evident that a cycle formed by $2n'$ vertices comprises exactly two 1-factors. If one assumes that the cycle C_{2n} represents the carbon skeleton of a completely conjugated hydrocarbon, then each of the 1-factors will correspond to one of the two Kekulé formulas of the ring system. This correspondence between the 1-factor and Kekulé formula also holds for other graphs representing the skeletons of completely conjugated hydrocarbons. In graph $\mathcal{G}_{25}$ there is no 1- factor, and so the corresponding compound, 2,3-dimethylenebutadiene-(1,3), cannot be represented by a Kekulé formula.

$$G_{25}:$$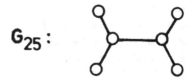

One can prove that all complete graphs $\mathcal{K}_{2n}$ can be 1-factorized. As an example, a 1-factorization of $\mathcal{K}_6 = \mathcal{G}_{20}$ is given below. Owing to the existence of six vertices in $\mathcal{K}_6$, any 1-factor consists of three

K_2. Consequently, the factors contain 3 edges. Because K_6 comprises altogether 15 edges, a 1-factorization of K_6 into 5 edge-disjoint factors exists as shown above. Chemists, familiar with the VB-method, identify the factors above as components of the valence bond-basis (i.e. all possible pairings of electrons with opposite spins) for benzene. The missing Kekulé components of the basis could be represented by two other 1-factors of K_6, which are particular linear combinations of the edge-disjoint 1-factors shown above.

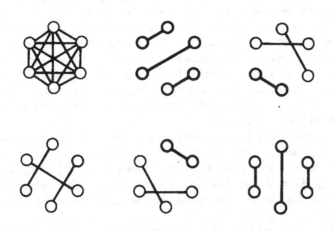

It is easy to show that for any $K_{2n'+1}$, n' edge-disjoint cycles of length $2n' + 1$ exist. Each of these $C_{2n'+1}$ is a 2-factor of $K_{2n'+1}$.

2.5 Planarity of Graphs

Before discussing this topic, an outline of two concepts is necessary.

A vertex x can be introduced into a graph G in such a way that x subdivides an edge $(a, b) \in G$ into two edges, namely (a, x) and (b, x). This procedure is called *edge subdivision*; the inverse operation is called *edge merging*. Vertices introduced into a graph G by edge subdivision have the degree two in G; in turn, only vertices of degree two can be removed from G by edge merging. The graphs G_{26} and G_{27} shown below are derived from $G_{18} = K_4$ by edge subdivision. Any edge subdivision will increase the number of vertices of degree two and enlarge $n(G)$ as well as $m(G)$ by one.

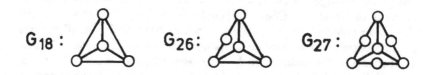

G_{18} : G_{26} : G_{27} :

Two graphs are called *homeomorphic* if they are derived from the same graph by edge subdivision. Graphs G_{26} and G_{27} are examples of homeomorphism; both graphs are homeomorphic to G_{18}. Since the conditions for homeomorphism are weaker than those for isomorphism, all isomorphic graphs are also homeomorphic. All cycles are homeomorphic and they are also homeomorphic to a graph consisting of one vertex and one loop. Moreover, all path graphs, P_n, are homeomorphic with K_2. In chemistry, the notion of homeomorphism has been converted into the concept of homologous series.

If a graph can be embedded in a plane or in the surface of a 3-dimensional sphere such that none of its edges intersects any other, the graph is called a *planar graph*. All the above graphs are planar, except for $K_5 = G_{19}$, $K_6 = G_{22}$ and the bipartite graph $K_{3,3} = G_{21} = G_{22}$, which are nonplanar. K_6 includes several unavoidable intersections of edges whereas K_5 and $K_{3,3}$ include only one.

The *theorem of Kuratowski* specifies both the necessary and sufficient conditions for the planarity of graphs: *A graph is planar if and only if it does not comprise subgraphs which are homeomorphic to K_5 or $K_{3,3}$.* The proof of this theorem, an outstanding achievement of mathematics, is not given here. In order to understand the meaning of this theorem the following facts are presented:

(i) K_5 and $K_{3,3}$ are the smallest nonplanar graphs with respect to the number of edges and vertices;

(ii) if the subgraph $G' \subset G$ is nonplanar, G cannot be planar;

(iii) only one unavoidable intersection of edges is sufficient in order to make G nonplanar.

The conjecture that constitutional graphs of chemical compounds are necessarily planar is false. The skeleton graph of the Möbius compound, synthesized by Walba *et al* in 1982, is homeomorphic to $K_{3,3}$.

Let G be a finite planar graph represented in a plane without any intersection of edges. The smallest independent cycles of G enclose by means of their vertices and edges definite parts of the plane. These parts of the plane are referred to as *interior faces* and are usually denoted as $\phi_1, \phi_2, \ldots$. The remaining part of the plane is called the *exterior face* ϕ_0. If there are no cycles in $\mathcal{G}$, the exterior face covers the whole plane. Thus, the set of faces of $\mathcal{G}$, $\mathcal{F} = \{\phi_0, \phi_1, \phi_2, \ldots\}$, is never empty.

The geometric dual $\mathcal{G}^*$ of a graph $\mathcal{G}$ is a graph obtained from $\mathcal{G}$ as follows:

(i) A vertex f_j^* is introduced into any face ϕ_j, $j = 0, 1, 2, \ldots$; the set of these vertices is the vertex set of $\mathcal{G}^*$, $\{f_j^*\} = \mathcal{V}(\mathcal{G}^*)$.

(ii) If faces ϕ_j and $\phi_{j'}$ of $\mathcal{G}$ meet at a common boundary comprising κ edges, the corresponding vertices f_j^* and $f_{j'}^*$ of $\mathcal{G}^*$ are to be connected by κ edges forming a κ-fold edge between f_j^* and $f_{j'}^* \in \mathcal{G}^*$.

This prescription for construction is illustrated using $\mathcal{G} = \mathcal{G}_{28}$. The vertices and edges of $\mathcal{G}_{28}$ are indicated by empty circles and full lines, and those of $\mathcal{G}_{28}^*$ by full circles and dashed lines, respectively. The geometric duals of $\mathcal{G}_{28}$ and $\mathcal{G}_{29}$ are isomorphic, $\mathcal{G}_{28}^* \cong \mathcal{G}_{29}$. Since the vertices of $\mathcal{G}_{29}$ are allowed to occupy arbitrary points in the plane in which they are drawn (they are not restricted to definite subsections), $\mathcal{G}_{29}$ is called the *dual* of $\mathcal{G}_{28}$. Note that in the geometric dual the location of edges and loops obey definite face relations, whereas such relations are not necessarily realized in the dual.

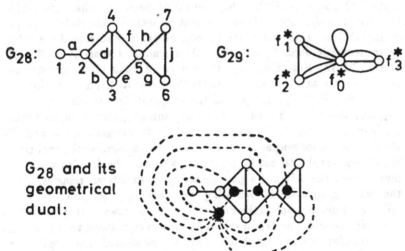

$\mathcal{G}_{28}$ and its geometrical dual:

The dual $\mathcal{T}^*$ of a tree $\mathcal{T}$ consists of a single vertex having the same number of loops as the edges $\mathcal{T}$ contains. The dual $\mathcal{C}_n^*$ of a cycle, $\mathcal{C}_n$, contains two vertices and one n-fold edge.

In general, the duals of simple graphs are multi- or pseudo- graphs; however, the dual of $\mathcal{K}_1 = \mathcal{G}_{15}$ is again $\mathcal{K}_1$ and thus $\mathcal{K}_1^* = \mathcal{K}_1$ (this is a selfdual graph).

The geometric dual $\mathcal{G}^*$ represents a definite mapping of vertices, edges and faces of a graph: The faces of $\mathcal{G}$ are mapped bijectively onto the vertices of $\mathcal{G}^*$, the vertices of $\mathcal{G}$ onto the faces of $\mathcal{G}^*$, and the edges of $\mathcal{G}$ onto the edges of $\mathcal{G}^*$:

$$\mathcal{V}(\mathcal{G}) \longleftrightarrow \mathcal{F}(\mathcal{G}^*), \quad \mathcal{E}(\mathcal{G}) \longleftrightarrow \mathcal{E}(\mathcal{G}^*), \quad \mathcal{F}(\mathcal{G}) \longleftrightarrow \mathcal{V}(\mathcal{G}^*). \quad (16)$$

From eq.(16) one can easily show that any planar simple graph is the geometric dual of its geometric dual:

$$(\mathcal{G}^*)^* = \mathcal{G}. \quad (17)$$

Thus, from its geometric dual $\mathcal{G}^*$, the original graph $\mathcal{G}$ can be regenerated. This statement, however, does not hold for the duals isomorphic with the geometric duals, as seen from the fact that $\mathcal{G}_{29}^* \neq \mathcal{G}_{28}$.

Subgraphs of duals, obtained by removing the vertex f_o^* corresponding to the exterior face of the original graph, were introduced into chemistry by Balaban *et al* to denote concisely the structure of polycyclic aromatic hydrocarbons (PAH); they are called *dualist* graphs. The graphs $\mathcal{G}_{30}$, $\mathcal{G}_{31}$, and $\mathcal{G}_{32}$ shown above represent respectively the skeleton graphs of naphthalene, anthracene, and phenanthrene. The scheme also exhibits their geometric duals $\mathcal{G}_j^*$ as well as their *truncated duals* $\mathcal{G}_j' = \mathcal{G}_j^* - f_0^*$, $j = 30, 31, 32$. Note that since the truncated duals $\mathcal{G}_j'$ are constructed in the same manner as the corresponding dualist graphs, $\mathcal{G}_j''$, they are isomorphic, $\mathcal{G}_j' \cong \mathcal{G}_j''$. In spite of the inequality of the graphs $\mathcal{G}_{31}$ and $\mathcal{G}_{32}$, and that of the geometric duals $\mathcal{G}_{31}^*$ and $\mathcal{G}_{32}^*$, their truncated duals are isomorphic, $\mathcal{G}_{31}' \cong \mathcal{G}_{32}'$ and, hence, their dualist graphs, $\mathcal{G}_{31}''$ and $\mathcal{G}_{32}''$ should also be isomorphic. To eliminate this ambiguity, the vertices of the dualist graphs cannot be placed anywhere in the drawing plane but must occupy the points of a grid consisting of equilateral triangles. Thus, the dualist graphs are derived using the following prescription:

(i) The skeleton graph of the PAH considered is drawn so that its vertices coincide with points of a grid consisting of equilateral triangles. This procedure guarantees that for any six-membered ring a grid point will be at its centre.

(ii) In the dualist graph of the PAH, $\mathcal{G}_j''$, the individual benzene rings are represented by vertices which are located at the central grid point of the ring. These vertices form the vertex set of $\mathcal{G}_j''$.

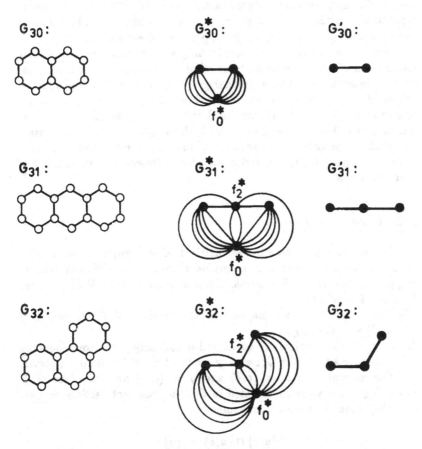

(iii) Two vertices of G_j'' are connected by an edge if the benzene rings to which they refer have an edge in common.

The prescriptions for the construction of G_j'' thus differ only in that the vertices of G_j'' must occupy grid points whereas those of G_j' may be arbitrarily located in the drawing plane. On the one hand this restriction with respect to the grid points is necessary in order to transfer the inequivalence of the skeleton graphs, $G_{31} \neq G_{32}$, to the corresponding dualist graphs, $G_{31}'' \neq G_{32}''$. On the other hand it introduces geometrical features into the concept of the dualist graph, namely the angles formed by adjacent edges. Since from the definition of graphs the dualist graphs are not graphs in the rigorous sense of this term. Nevertheless, they are very useful and, indeed, the dualist graphs of the PAH's not only provide a practical shorthand for the structure of these compounds, but are also valuable tools in correlation analysis, polynomial expansions, etc.

Finally, we mention that nonplanar graphs having one unavoidable intersection of edges can be represented on the surface of a torus without such an intersection. The nonplanar graph $K_{3,3}$, as well as the skeleton graph of the Möbius compound synthesized by Walba *et al* (which is homeomorphic to $K_{3,3}$) can be drawn on a torus surface without intersections. The surface of the torus is also important for other unusual chemical compounds: The skeletons of catenanes, rotaxanes and knots cannot be embedded in the surface of a three-dimensional sphere, but in that of a torus.

2.6 Line Graphs

In the previous section the geometric dual of a graph was introduced as a kind of a bijective nonisomorphic mapping. A different kind of mapping produces the *line graph*, $\mathcal{L}(\mathcal{G})$, of a graph $\mathcal{G} = [\mathcal{V}, \mathcal{E}]$ and may be derived as follows:

(i) The edges of $\mathcal{G}$ are mapped as the vertices of $\mathcal{L}(\mathcal{G})$, and thus $\mathcal{E}(\mathcal{G}) \longleftrightarrow \mathcal{V}(\mathcal{L}(\mathcal{G}))$.

(ii) In $\mathcal{L}(\mathcal{G})$ two vertices are connected if and only if the corresponding edges are adjacent in $\mathcal{G}$, i.e. they are incident with a common vertex.

For example, let $x = (a, b)$ and $y = (a, c)$ be two edges of $\mathcal{G}$, $x, y \in \mathcal{E}(\mathcal{G})$. They are represented in $\mathcal{L}(\mathcal{G})$ by adjacent vertices because in $\mathcal{G}$ they have the vertex a in common:

$$\{a, b\} \cap \{a, c\} = \{a\}.$$

Because of this relation, the line graph $\mathcal{L}(\mathcal{G})$ is sometimes also called an *intersection graph*; yet another name, the *interchange graph*, refers to the bijective mapping $\mathcal{E}(\mathcal{G}) \longleftrightarrow \mathcal{V}(\mathcal{L}(\mathcal{G}))$. But it should be noted that the term "intersection graph" has a more general meaning and that line graphs represent only special cases of intersection graphs. As an illustration, $\mathcal{G}_{28}$ and its line graph, $\mathcal{G}_{33} = \mathcal{L}(\mathcal{G}_{28})$, are illustrated below. The letters a to j correlate the edges of $\mathcal{G}_{28}$ with the vertices of $\mathcal{G}_{33}$.

As a consequence of the way in which the line graph is constructed, the edge wreath $\mathcal{W}(v_j)$ of the vertices $v_j \in \mathcal{G}$ is mapped as a complete graph $\mathcal{L}(\mathcal{W}(v_j)) = \mathcal{K}_g$ in the line graph $\mathcal{L}(\mathcal{G})$, where $g = g(v_j)$ designates

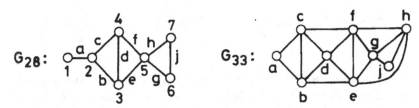

the degree of the vertex $v_j \in \mathcal{G}$. The line graph $\mathcal{L}(\mathcal{G})$ is the union of all the $\mathcal{L}(\mathcal{W}(v_j))$, where all vertices bearing the same label are identified as one vertex. Hence, it follows that line graphs are planar only if in the parent graph no vertex has degree $g > 4$.

If two graphs are isomorphic, $\mathcal{G}' \cong \mathcal{G}''$, then their line graphs are also isomorphic, $\mathcal{L}(\mathcal{G}') \cong \mathcal{L}(\mathcal{G}'')$. The converse statement has a single exception, namely that the line graphs of the nonisomorphic graphs $\mathcal{K}_3 = \mathcal{G}_{17}$ and $\mathcal{K}_{1,3} = \mathcal{G}_{34}$ are isomorphic, $\mathcal{L}(\mathcal{K}_3) \cong \mathcal{L}(\mathcal{K}_{1,3}) \cong \mathcal{K}_3$.

At present, the importance of line graphs in chemistry has not been adequately recognized. The vertices of the line graphs of constitutional graphs may be thought of as the images of the localized σ-bonds of the molecule considered. Thus, line graphs map a particular σ-basis in a manner similar to that for π-AO's mapped by skeleton graphs. Consequently, line graphs furnish a tool for studying the topological properties of σ-systems (see Section 13).

2.7 Operations on Graphs

Operations on graphs are those rules employed to form a new graph $\mathcal{H} = \mathcal{H}(\mathcal{H}_1, \mathcal{H}_2)$ from two graphs, $\mathcal{H}_1 = [\mathcal{V}_1, \mathcal{E}_1]$ and $\mathcal{H}_2 = [\mathcal{V}_2, \mathcal{E}_2]$. In general, operations on graphs are commutative, $\mathcal{H}(\mathcal{H}_1, \mathcal{H}_2) = \mathcal{H}(\mathcal{H}_2, \mathcal{H}_1)$; exceptions will be noted explicitly. The results arising from the different operations on graphs are illustrated in terms of $\mathcal{H}_1 = \mathcal{K}_2$ and $\mathcal{H}_2 = \mathcal{P}_3$.

$$H_1 = K_2 : \quad \underset{a \quad b}{\circ\!\!-\!\!\circ} \qquad\qquad H_2 = P_3 : \quad \underset{1 \quad 2 \quad 3}{\circ\!\!-\!\!\circ\!\!-\!\!\circ}$$

The *union* $\mathcal{H} = \mathcal{H}_1 \cup \mathcal{H}_2$ is defined via the union of the edge and vertex sets: $V(\mathcal{H}) = V_1 \cup V_2$, $\mathcal{E}(\mathcal{H}) = \mathcal{E}_1 \cup \mathcal{E}_2$. If $\mathcal{H}_1$ and $\mathcal{H}_2$ are disjoint connected graphs, then $\mathcal{H}$ consists of the two components $\mathcal{H}_1$ and $\mathcal{H}_2$; if, however, $\mathcal{H}_1$ and $\mathcal{H}_2$ are two connected graphs but not disjoint, then $\mathcal{H}$ is connected (see Section 6).

$$H = K_2 \cup P_3 :$$

A *join* $\mathcal{H} = \mathcal{H}_1 \oplus \mathcal{H}_2$ differs from the union $\mathcal{H}_1 \cup \mathcal{H}_2$ only in that all vertices of $\mathcal{H}_1$ and $\mathcal{H}_2$ are interconnected.

$$\mathcal{H} = \mathcal{H}_1 \oplus \mathcal{H}_2 = \mathcal{H}_1 \cup \mathcal{H}_2 \cup \{(x_\xi, y_\eta)\}$$

$$x_\xi \in \mathcal{H}_1, \quad 1 \leq \xi \leq n_1,$$

$$y_\eta \in \mathcal{H}_2, \quad 1 \leq \eta \leq n_2.$$

$$H = K_2 \oplus P_3 :$$

Any complete bipartite graph K_{n_1,n_2} is a join of two completely unconnected (edgeless) graphs ($m_1 = 0$, $m_2 = 0$).

The *(Cartesian) product* $\mathcal{H} = \mathcal{H}_1 \otimes \mathcal{H}_2$ is defined upon the Cartesian product of the two vertex sets of both graphs; $V(\mathcal{H}) = V_1 \otimes V_2 = \{x_\xi y_\eta | 1 \le \xi \le n_1, \quad 1 \le \eta \le n_2\}$; in the cases of $\mathcal{H}_1 = K_2$ and $\mathcal{H}_2 = P_3$, $V(K_2 \otimes P_3) = \{1a, 1b, 2a, 2b, 3a, 3b\}$. Two vertices $x_\xi y_\eta$ and $x_\zeta y_\omega$ are adjacent in $\mathcal{H}$ if and only if either $x_\xi = x_\zeta$ and the edge $(y_\eta, Y_\omega) \in \mathcal{H}_2$ or $y_\eta = y_\omega$ and the edge $(x_\xi, x_\zeta) \in \mathcal{H}_1$.

$$H = K_2 \otimes P_3 :$$

A noncommutative operation, also defined via the cartesian product $V_1 \otimes V_2$ is the *composition* of $\mathcal{H} = \mathcal{H}_1[\mathcal{H}_2]$. In $\mathcal{H}$ the two vertices $x_\xi y_\eta$ and $x_\zeta y_\omega$ are connected if and only if either in $\mathcal{H}_1$ the edge of $(x_\xi, x_\zeta) \in \mathcal{H}_1$ exists or $x_\xi = x_\zeta$ and in $\mathcal{H}_2$ the edge $(y_\eta, y_\omega) \in \mathcal{H}_2$ occurs; in the case of $(x_\xi, x_\zeta) \in \mathcal{H}_1$, the interrelation of y_η and y_ω is trivial. Since the edge relationships in $\mathcal{H}_1$ and $\mathcal{H}_2$ influence the edge set $\mathcal{E}(\mathcal{H})$ in different ways, the two compositions, derived from $\mathcal{H}_1$ and $\mathcal{H}_2$ $(\mathcal{H}_1 \ne \mathcal{H}_2)$ are nonisomorphic, i.e. $\mathcal{H}_2[\mathcal{H}_1] \ne \mathcal{H}_1[\mathcal{H}_2]$. That implies that composition is not a commutative graph operation.

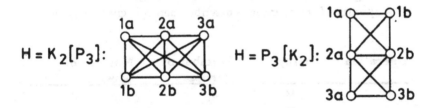

$$H = K_2[P_3]: \qquad H = P_3[K_2]:$$

The *corona* $\mathcal{H} = \mathcal{H}_1 \circ \mathcal{H}_2$ is defined on the union of $\mathcal{H}_1$ with n_1 copies of $\mathcal{H}_2$, where n_1 denotes the number of vertices of $\mathcal{H}_1$. When the j-th vertex of $\mathcal{H}_1$ is connected with all vertices of the j-th copy of $\mathcal{H}_2$ and this procedure is performed for all vertices of $\mathcal{H}_1$, $j = 1, 2, \ldots, n_1$, the corona will be formed. The vertex set of the corona $\mathcal{H} = \mathcal{H}_1 \circ \mathcal{H}_2$ is hence the union of V_1 and the Cartesian product $V_1 \otimes V_2$:

$$V(\mathcal{H}_1 \circ \mathcal{H}_2) = V_1 \cup (V_1 \otimes V_2).$$

This way of constructing the vertex set implies that from two different graphs, $\mathcal{H}_1 \neq \mathcal{H}_2$, two nonisomorphic coronas will be formed, $\mathcal{H}_1 \circ \mathcal{H}_2 \neq \mathcal{H}_2 \circ \mathcal{H}_1$.

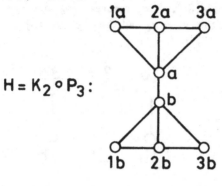

$H = K_2 \circ P_3:$

$H = P_3 \circ K_2:$

In addition to the operations on graphs outlined above, a number of further operations exist, whose significance in chemistry is not apparent at present. The graphs we have introduced are occasionally employed in the investigation of problems in chemical reactivity.

A synopsis of the operations described herein is given in Table 1.

Operation	Graph	Number of vertices	Number of edges
	$\mathcal{H}_1$	n_1	m_1
	$\mathcal{H}_2$	n_2	m_2
Union	$\mathcal{H}_1 \cup \mathcal{H}_2$	$n_1 + n_2$	$m_1 + m_2$
Join	$\mathcal{H}_1 \oplus \mathcal{H}_2$	$n_1 + n_2$	$m_1 + m_2 + n_1 n_2$
Product	$\mathcal{H}_1 \otimes \mathcal{H}_2$	$n_1 n_2$	$n_1 m_2 + n_2 m_1$
Composition	$\mathcal{H}_1[\mathcal{H}_2]$	$n_1 n_2$	$n_1 m_2 + n_2^2 m_1$
Corona	$\mathcal{H}_1 \circ \mathcal{H}_2$	$n_1(n_2 + 1)$	$m_1 + n_1(n_2 + m_2)$

Operations defined upon two graphs $\mathcal{H}_1$ and $\mathcal{H}_2$

No.	Symbol	Permutation	Cycle structure	Number of transpositions
1	E	$\begin{pmatrix}1234567\\1234567\end{pmatrix}$	$[1]^7$	0
2	P	$\begin{pmatrix}1234567\\1243567\end{pmatrix}$	$[1]^5[2]$	1
3	Q	$\begin{pmatrix}1234567\\1234576\end{pmatrix}$	$[1]^5[2]$	1
4	R	$\begin{pmatrix}1234567\\1243576\end{pmatrix}$	$[1]^3[2]^2$	2

Automorphisms of $\mathcal{G}_{28}$

2.8 The Automorphism Group of a Graph

An isomorphic mapping of a graph $\mathcal{G}$ onto itself is called an *automorphism*. Thus, an automorphism is a bijective mapping of $V(\mathcal{G})$ onto itself, which maintains the pair relationship $\mathcal{E}(\mathcal{G})$:

$$V(\mathcal{G}) \longleftrightarrow V(\mathcal{G}), \quad \mathcal{E}(\mathcal{G}) \longleftrightarrow \mathcal{E}(\mathcal{G}). \tag{18}$$

Any graph has at least one such automorphism, namely the *identity* mapping, where any vertex or edge is mapped onto itself:

$$v_j \longleftrightarrow v_j, \quad v_j \in V(\mathcal{G});$$

$$e_k \longleftrightarrow e_k, \quad e_k \in \mathcal{E}(\mathcal{G}). \tag{19}$$

The occurrence of further automorphisms, in addition to the identity mapping, will depend on the symmetry of the graph.

By inspection, the following automorphisms of $\mathcal{G}_{28}$ are seen to exist:
(i) identity mapping;
(ii) mutual mapping of the vertices 3 and 4 onto each other, $3 \longleftrightarrow 4$;
(iii) mutual mapping of the vertices 6 and 7 onto each other, $6 \longleftrightarrow 7$;
(iv) simultaneous mutual mappings of $3 \longleftrightarrow 4$ and $6 \longleftrightarrow 7$.

The vertices not mentioned explicitly above are mapped onto themselves. In Table 2 the automorphic mappings of $\mathcal{G}_{28}$ are properly represented by means of the permutations of the vertex numbers; the upper line designates the original numbering whereas the lower line refers to the images. To maintain concise notation, the different permutations are symbolized by E, P, Q, R; the identical permutation is usually denoted by E or I.

Seven vertices give rise to 7! = 5040 mappings onto themselves; however, only the four mappings given in Table 2 are automorphic mappings of $\mathcal{G}_{28}$. This is a consequence of the requirement that the pair relationships must be preserved, i.e. only *equivalent vertices* can be mapped automorphically onto each other. Equivalent vertices form disjoint subsets of $\mathcal{V}(\mathcal{G})$ called *orbits*. Obviously, the symmetry of a graph is reflected by the partitioning of its vertex set into orbits. For $\mathcal{G}_{28}$ the following orbits are obtained:

$$\{1\}, \quad \{2\}, \quad \{3,4\}, \quad \{5\}, \quad \{6,7\}. \qquad (20)$$

As may be seen, vertices 1, 2 and 5 can be mapped automorphically only onto themselves, whereas 3 may be mapped onto 4 and vice versa; the analogue holds for vertices 6 and 7. In $\mathcal{G}_{28}$ the mappings 3 $\longleftrightarrow$ 4 and 6 $\longleftrightarrow$ 7 can be performed independently; in graph $\mathcal{G}_{35} = \mathcal{G}_{28} \cup \{(3,6), (4,7)\}$, however, it is necessary to perform them simultaneously. Accordingly, only E and R are automorphisms of $\mathcal{G}_{35}$.

$$G_{35}:$$

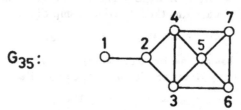

The permutations given in Table 2 can be more concisely represented as follows:

$$E = \begin{pmatrix} 1 \\ 1 \end{pmatrix} \begin{pmatrix} 2 \\ 2 \end{pmatrix} \begin{pmatrix} 3 \\ 3 \end{pmatrix} \begin{pmatrix} 4 \\ 4 \end{pmatrix} \begin{pmatrix} 5 \\ 5 \end{pmatrix} \begin{pmatrix} 6 \\ 6 \end{pmatrix} \begin{pmatrix} 7 \\ 7 \end{pmatrix},$$

$$P = \begin{pmatrix} 1 \\ 1 \end{pmatrix} \begin{pmatrix} 2 \\ 2 \end{pmatrix} \begin{pmatrix} 34 \\ 43 \end{pmatrix} \begin{pmatrix} 5 \\ 5 \end{pmatrix} \begin{pmatrix} 6 \\ 6 \end{pmatrix} \begin{pmatrix} 7 \\ 7 \end{pmatrix},$$

$$Q = \begin{pmatrix} 1 \\ 1 \end{pmatrix} \begin{pmatrix} 2 \\ 2 \end{pmatrix} \begin{pmatrix} 3 \\ 3 \end{pmatrix} \begin{pmatrix} 4 \\ 4 \end{pmatrix} \begin{pmatrix} 5 \\ 5 \end{pmatrix} \begin{pmatrix} 67 \\ 76 \end{pmatrix},$$

$$R = \begin{pmatrix} 1 \\ 1 \end{pmatrix} \begin{pmatrix} 2 \\ 2 \end{pmatrix} \begin{pmatrix} 34 \\ 43 \end{pmatrix} \begin{pmatrix} 5 \\ 5 \end{pmatrix} \begin{pmatrix} 67 \\ 76 \end{pmatrix}. \qquad (21)$$

In this notation transposed vertices are assembled into *cycles* to emphasize the *cycle structure* of the corresponding permutation. Thus, E consists of

1-cycles only, P and Q are composed of one 2-cycle and five 1-cycles each, while R has two 2-cycles and three 1-cycles (see Table 2).

Of course, in permutations 3-cycles, 4-cycles, etc. may also occur. For example, in the automorphisms of the complete graph $\mathcal{K}_n$ all m-cycles $(1 \leq m \leq n)$ occur. By means of the cycle structure, any permutation can be indicated in one line, such as $R = (1)\,(2)\,(34)\,(5)\,(67)$.

A mutual interchange of two elements is called a *transposition*. A 2-cycle therefore corresponds with a transposition. It can be shown that any m-cycle is generated by m-1 transpositions. An *even (odd) permutation* is produced by an even (odd) number of transpositions. Thus, $[2\nu]$ indicates an odd but $[2\nu{+}1]$ an even permutation cycle. If a particular transposition is performed twice and consecutively, the original sequence of objects is regenerated. Therefore, from eq.(21) it follows that:

$$P^2 = E, \quad Q^2 = E, \quad R^2 = E. \tag{22}$$

In the permutation R the transpositions (34) and (67) are performed simultaneously whereas in P and Q these transpositions are executed separately. R may thus be expressed as:

$$PQ = QP = R. \tag{23}$$

From eqs.(22) and (23), it is evident that the permutations listed in Table 2 form a group. This group is called the *automorphism group* $\mathcal{A}(\mathcal{G}_{28})$ of the graph $\mathcal{G}_{28}$, i.e.

$$\mathcal{A}(\mathcal{G}_{28}) = \{E, P, Q, R\}. \tag{24}$$

A nonempty set $\mathcal{A} = \{A_j | j = 1, 2, \dots\}$ on which a binary relation (symbolized by $\odot$) is defined, for instance $A_i \odot A_j = A_k$ (concisely denoted by $A_i A_j = A_k$), is said to form a group under the binary relation law if the following axioms are satisfied:

[G 1] *Closure property:* The product of any two elements in the set is also an element of the set, i.e. if $A_1, A_2 \in \mathcal{A}$ and $A_1 A_2 = A_3$ then $A_3 \in \mathcal{A}$.

[G 2] *The associative law* holds for any triple product of the elements:

$$A_1(A_2 A_3) = (A_1 A_2)A_3. \tag{25}$$

[G 3] *Identity element:* The group $\mathcal{A}$ must contain the identity element E which satisfies the relation:

$$E\,A_j = A_j\,E = A_j, \tag{26}$$

where A_j is any element of the group, $A_j \in \mathcal{A}$.

[G 4] *Inverse elements*: Every element A_j must have an inverse, $A_j^{-1} = A_k$ such that

$$A_j A_k = A_k A_j = E. \tag{27}$$

From eqs. (26) and (27) it follows that $E^{-1} = E$.

The elements of the group may or may not commute, i.e. $A_j A_k = A_k A_j$, or $A_j A_k \neq A_k A_j$. The identity element, E, always commutes with any element of the group according to eq.(26). When all elements of a group commute the group is called *Abelian*; for instance, $\mathcal{A}(\mathcal{G}_{28})$ is an Abelian group.

The number of elements of a group is said to be its *order*, $h(\mathcal{A})$. As seen from eq.(24), the order of $\mathcal{A}(\mathcal{G}_{28})$ is $h(\mathcal{A}(\mathcal{G}_{28})) = 4$.

The number of permuted objects, on which the elements of the group operate, is called the *degree* of the group, $g(\mathcal{A})$; evidently, in general $g(\mathcal{A}(\mathcal{G})) = n(\mathcal{G})$ holds. In the case of $\mathcal{G}_{28}$, $g(\mathcal{A}(\mathcal{G}_{28})) = 7$.

For permutation groups the degree is a characteristic property which is not defined in point groups. From Table 2 it can easily be verified that for any permutation of the cycle structure $[1]^{\alpha_1}[2]^{\alpha_2}[3]^{\alpha_3} \cdots = \prod[\kappa]^{\alpha_\kappa}$ the following relationship holds:

$$g(\mathcal{A}) = \sum_{\kappa=1}^{n} \kappa \, \alpha_\kappa. \tag{28}$$

As in point groups, the elements of the automorphism groups form *classes* generated by performing the operation:

$$X \, B \, X^{-1} = C(B,X)$$

for all elements $X \in \mathcal{A}$. The set $\{C(B,X) \mid X \in \mathcal{A}\}$ represents a class which also includes B; this inclusion of B is guaranteed at least by $EBE^{-1} = B$. Each element of $\mathcal{A}$ belongs to one and only one class. Eqs.(26) and (27) show that E forms a class on its own. It can be established that two elements belonging to the same class have the same cycle structure; the converse, however, is not true.

As in point groups, the number of classes equals the number of *irreducible representations* $\{\Gamma_j\}$ of the group $\mathcal{A}$. In Abelian groups any element forms a class on its own. Consequently, in Abelian groups the number of classes is identical to their order. Abelian groups thus possess $h(\mathcal{A})$ one-dimensional irreducible representations. Two and multidimensional irreducible representations are referred to as *doubly* or *multi-degenerate*. For further properties of representations reference should be made to the extensive literature on group theory.

All the products of the elements of $\mathcal{A}(\mathcal{G}_{28})$ follow from eqs.(22) and (23); they are summarized in Table 3, which for historical reasons is known as the *multiplication table* of the group $\mathcal{A}(\mathcal{G}_{28})$. Using this table the group axioms can be verified.

	E	P	Q	R	X	X^{-1}
E	E	P	Q	R	E	E
P	P	E	R	Q	P	P
Q	Q	R	E	P	Q	Q
R	R	Q	P	E	R	R

Multiplication Table and Inverse Elements of $\mathcal{A}(\mathcal{G}_{28})$

Since $\mathcal{A}(\mathcal{G}_{28})$ is an Abelian group, any element is a class on its own. Consequently, $\mathcal{A}(\mathcal{G}_{28})$ has four irreducible representations, denoted as Γ_0 to Γ_3. The characters of the elements of $\mathcal{A}(\mathcal{G}_{28})$ in the individual irreducible representations are given in Table 4, which is the *character table* of the group $\mathcal{A}(\mathcal{G}_{28})$. For determining the characters $X(A_j, \Gamma_k)$ reference should again be made to the literature on group theory.

	E	P	Q	R
Γ_0	1	1	1	1
Γ_1	1	1	-1	-1
Γ_2	1	-1	1	-1
Γ_3	1	-1	-1	1

Character Table of $\mathcal{A}(\mathcal{G}_{28})$

When the elements of two groups can be mapped bijectively onto each other, such that the binary relations expressed in their multiplication tables are maintained, the two groups are said to be isomorphic. As a consequence, *isomorphic groups* have identical character tables. Both these groups are isomorphic with a particular *abstract group* consisting of elements without a concrete meaning. In the case of $A(\mathcal{G}_{28})$, an isomorphism with the point groups C_{2v}, C_{2h} and D_2 exists. A geometric arrangement of seven points according to $\mathcal{G}_{28}$ can exhibit no higher symmetry than $C_{2v} = \{E, C_2, \sigma_v, \sigma_v'\}$. This symmetry can be achieved by two different configurations (see Figure 1): the points corresponding to the vertices $\{3,4\}$ and $\{6,7\}$ are located either in different planes (σ_v and σ_v' see Figure 1a) or in the same plane (σ_v see Figure 1b). In the first case, any permutation of $A(\mathcal{G}_{28})$ corresponds to a particular symmetry element of C_{2v}. In the second case, each of the permutations $E, R \in A(\mathcal{G}_{28})$ is attributed to two symmetry elements, namely $E, \sigma_v \in C_{2v}$ and $C_2, \sigma_v' \in C_{2v'}$, respectively. However, the permutations $P, Q \in A(\mathcal{G}_{28})$ find no equivalent in the symmetry elements of C_{2v}; they represent *local symmetries* which cannot be accounted for by the symmetry elements of point groups because their symmetry elements act simultaneously on all points of the space. Local symmetries, however, are very important in the discussion of the symmetries of *nonrigid molecules* which are being increasingly studied in a variety of chemical problems, especially those involving use of NMR spectroscopy. The automorphism group of the constitutional graph of the compound considered satisfies the constraints imposed by these fields.

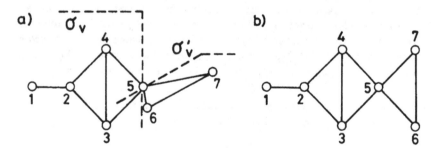

Two configurations of the diagram corresponding to $\mathcal{G}_{28}$. The configurations exhibit the highest possible symmetry, C_{2v}.

The permutation groups of degree n used hereafter are given in Table 5. The *symmetric group* S_n is the automorphism group of the complete graph, $\mathcal{K}_n$, as well as that of its complement, $\overline{\mathcal{K}}_n$, the completely unconnected graph having n isolated vertices.

The *dihedral group*, $\mathcal{D}_n$, is the automorphism group of the cycle, C_n,

Name	Order	Description
S_n symmetric	$n!$	all permutations of $\{1, 2, \ldots, n\}$
A_n alternating	$n!/2$	all even permutations of $\{1, 2, \ldots, n\}$
$\mathcal{D}_n$ dihedral	$2n$	generated by the n-cycle $(1, 2, \ldots, n)$ and $(1n)(2n-1)(3n-2)\ldots$
C_n cyclic	n	generated by the cycle $(1, 2, \ldots, n)$
$\mathcal{E}_n$ identity	1	contains only the identical permutation $(1)(2)\ldots(n)$

and that of its complement $\overline{C}_n = K_n \setminus C_n$:

$$A(K_n) = A(\overline{K}_n) = S_n \quad A(C_n) = A(\overline{C}_n) = \mathcal{D}_n. \tag{29}$$

All permutation groups of degree n and of order less than $n!$ are proper subgroups of S_n (*Cayley's theorem*).

The graph operations outlined in Section 7 have their counterpart in the *operations on automorphism groups*. In order to explain the concept of *group operations*, two permutation groups, $A = \{A_j\}$ and $B = \{B_k\}$, will be used; they may be considered as the automorphism groups of two graphs. A and B act on the object sets $\mathcal{X} = \{x_\xi\}$ and $\mathcal{Y} = \{y_\eta\}$, respectively, such that:

$$(A_j x_\xi) \in \mathcal{X}, \qquad (B_k y_\eta) \in \mathcal{Y}, \tag{30}$$

i.e. the action of an element of these groups on an element of its object set results in an element of this set. The orders and the degrees of these groups may be denoted as follows:

$$h(A) = p, \qquad h(B) = q;$$
$$g(A) = |\mathcal{X}| = d, \qquad g(B) = |\mathcal{Y}| = e.$$

The *sum* or the *direct product* $A \oplus B$ is a permutation group operating on the union set $Z = \mathcal{X} \cup \mathcal{Y}$. Consequently, the degree of this permutation group is $g(A \oplus B) = d + e$ and its elements are ordered pairs $(A_j \oplus B_j)$ from which $h(A \oplus B) = pq$ follows. The elements $z \in Z$ are permuted using the following rule:

$$(A_j \oplus B_k)z = \begin{cases} = (A_j z) \ldots & \text{if } z \in \mathcal{X}, \\ = (B_k z) \ldots & \text{if } z \in \mathcal{Y}. \end{cases} \tag{31}$$

The direct product corresponds to two operations on graphs, i.e. the union $\mathcal{H} = \mathcal{H}_1 \cup \mathcal{H}_2$ and the join $\mathcal{H} = \mathcal{H}_1 \oplus \mathcal{H}_2$. Consequently, for the corresponding automorphism group one obtains:

$$A(\mathcal{H}_1 \cup \mathcal{H}_2) = A(\mathcal{H}_1) \oplus A(\mathcal{H}_2);$$
$$A(\mathcal{H}_1 \oplus \mathcal{H}_2) = A(\mathcal{H}_1) \oplus A(\mathcal{H}_2). \tag{32}$$

Complete bipartite graphs can be thought of as a join of two completely unconnected graphs where $K_{n',n''} = \overline{K}_{n'} \oplus \overline{K}_{n''}$. From eqs.(29) and (32) for the automorphism group one derives the relationship:

$$A(K_{n',n''}) = S_{n'} \oplus S_{n''}. \tag{33}$$

The *Cartesian product* $A \otimes B$ is a permutation group which operates on the Cartesian product of the sets, $\mathcal{Z} = \mathcal{X} \otimes \mathcal{Y}$. $\mathcal{Z}$ is the set $z_{\xi\eta} = x_\xi \otimes y_\eta$, $1 \leq \xi = |\mathcal{X}| = d$, and $1 \leq \eta \leq |\mathcal{Y}| = e$; hence, $g(A \otimes B) = |\mathcal{Z}| = de$. Again, the elements of $A \otimes B$ are ordered pairs, $(A_j \otimes B_k)$, and so the order will be $h(A \otimes B) = pq$. From a formal point of view, the number of elements and the mode of their construction is the same as that for the groups $(A \oplus B)$ and $(A \otimes B)$. However, the object on which they operate, as well as the results of these operations, are different. Unlike eq.(31), the element $z_{\xi\eta} = x_\xi \otimes y_\eta$ is permuted as follows:

$$(A_j \otimes B_k)(x_\xi \otimes y_\eta) = (A_j \otimes x_\xi) \otimes (B_k \otimes y_\eta). \tag{34}$$

The Cartesian product $A \otimes B$ corresponds to the Cartesian product of the two graphs $\mathcal{H} = \mathcal{H}_1 \otimes \mathcal{H}_2$. Provided $A(\mathcal{H}_1)$ and $A(\mathcal{H}_2)$ are the groups given in Table 5, the automorphism group $A(\mathcal{H})$ will take the following form:

$$A(\mathcal{H}_1 \otimes \mathcal{H}_2) = A(\mathcal{H}_1) \otimes A(\mathcal{H}_2). \tag{35}$$

The *wreath product*, $A[B]$, also operates on the set $\mathcal{Z} = \mathcal{X} \otimes \mathcal{Y}$; therefore $g(A[B]) = de$ is again true. For any $A_j \in A$ and any sequence $(B_1, B_2, \ldots B_d)$ of d permutations of B which are not necessarily different, a permutation $(A_j; B_1, \ldots B_d)$ of $A[B]$ exists, with the element $z_{\xi\eta} = x_\xi \otimes y_\eta$ permuted as follows:

$$(A_j; B_1, B_2, \ldots, B_d)(x_\xi \otimes y_\eta) = (A_j x_\xi) \otimes (B_\xi y_\eta). \tag{36}$$

From the mode of constructing the permutations of $A[B]$ it follows that the order $h(A[B]) = pq^d$. Because the permutations of the outer group,

A, and those of the inner group, B, are used in a different manner for the construction of the elements of $A[B]$, the two possible wreath products, $A[B]$ and $B[A]$ are nonisomorphic; i.e. $A[B] \neq B[A]$. The wreath product is of major importance in discussions of the symmetry of nonrigid molecules or regular polymers.

The *power group* B^A is a less frequently occurring permutation group. It operates on the set y^X, the set of all mappings of X into y. Consequently, the degree of the power group is $g(B^A) = e^d$. For any pair $A_j \in A$ and $B_k \in B$, a permutation $B_k^{A_j}$ exists in B^A whose operation on a mapping f from y^X is defined as follows:

$$(B_k^{A_j} f)(x) = B_k(fA_j x)). \tag{37}$$

This equation indicates for any $x \in X$ the image which is obtained by applying the mapping $B_k^{A_j} f$ on x.

	Group	Object Set	Order	Degree
	A	X	p	d
	B	y	q	e
Direct product	$A \oplus B$	$X \cup y$	pq	$d + e$
Cartesian product	$A \otimes B$	$X \otimes y$	pq	de
Wreath product	$A[B]$	$X \otimes y$	pq^d	de
Power group	B^A	y^X	pq	e^d

Groups Composed from Two Permutation Groups A and B

In Table 6 the above mentioned groups constructed from two permutation groups and their characteristics are summarized.

One can show that the direct product, the Cartesian product, and the power group are all isomorphic, i.e. that:

$$A \oplus B \cong A \otimes B \cong B^A, \tag{38}$$

despite the fact that these groups have different degrees. Moreover, it can also be shown that in these three composed groups the groups A and B commute such that:

$$A \oplus B \equiv B \oplus A, \quad A \otimes B \equiv B \otimes A;$$
$$B^A \cong A^B. \tag{39}$$

In the last part of this section the way in which the automorphism group of a graph, say $A(\mathcal{G}_{28})$, is constructed using the permutation groups given in Table 5 is outlined. As illustrated in Table 2, the vertices $1, 2, 5 \in \mathcal{G}_{28}$ can only be mapped identically. These mappings form a factor of $A(\mathcal{G}_{28})$ which is the identity group $\mathcal{E}_3$ operating on the vertices, 1, 2 and 3. The vertices of each of the pairs 3,4 and 6,7 can be mapped either identically or on to each other. The respective permutations form two symmetric groups $\mathcal{S}_2$ operating on the pairs, 3,4 and 6,7, respectively. The group $A(\mathcal{G}_{28})$ acts on the union of the vertex sets on which these three factors operate. Thus, one may expect that:

$$A(\mathcal{G}_{28}) = \mathcal{E}_3 \oplus \mathcal{S}_2 \oplus \mathcal{S}_2. \tag{40}$$

Using the entries of Tables 5 and 6, it follows that $g(A(\mathcal{G}_{28})) = 3+2+2 = 7$ and $h(A(\mathcal{G}_{28})) = 1 \cdot 2 \cdot 2 = 4$, which accords with the results in Table 2. A final check for the validity of eq.(40) is obtained when all permutations according to eq.(40) are set up explicitly and then compared with those of Table 2.

The automorphism group of the graph $\mathcal{G}_{35} = \mathcal{G}_{28} \cup \{(3,6),(4,7)\}$ comprises again $\mathcal{E}_3$ as a factor which operates on the vertices 1, 2 and 5. The vertices 3 and 6 of $\mathcal{G}_{35}$, however, can be mapped only as a pair onto the pair of vertices 4 and 7 because both pairs are connected by an edge. Thus, a factor of $A(\mathcal{G}_{35})$ acts on these four vertices and this can be formulated as the wreath product $\mathcal{S}_2[\mathcal{E}_2]$. One therefore obtains for $A(\mathcal{G}_{35})$ the result:

$$A(\mathcal{G}_{35}) = \mathcal{E}_3 \oplus (\mathcal{S}_2[\mathcal{E}_2]). \tag{41}$$

Using Tables 5 and 6 it follows that $g(A(\mathcal{G}_{35})) = 3 + 2 \cdot 2 = 7$ and $h(A(\mathcal{G}_{35})) = 1 \cdot (2 \cdot 1^2) = 2$.

Representative examples for permutation groups describing the *symmetries of nonrigid molecules* are the automorphism groups of the constitutional graphs for boron trifluoride, BF_3, and trimethylborane, $B(CH_3)_3$. Evidently, for BF_3 one derives the expression:

$$A(BF_3) = \mathcal{E}_1 \oplus \mathcal{S}_3. \tag{42}$$

Here $\mathcal{E}_1$ operates on the vertex representing the boron atom, whereas $\mathcal{S}_3$ acts on the vertices corresponding to the fluorine atoms. In $B(CH_3)_3$, the F atoms are replaced by CH_3 groups for which $\mathcal{E}_1 \oplus \mathcal{S}_3$ is derived when a separated methyl group is considered. If a methyl group is visualized as a single entity, any methyl group can be mapped onto any other methyl group. Thus, the methyl entities behave like the fluorine atoms in BF_3. Consequently, the automorphism group corresponding to trimethylborane has the same global structure as that given by eq.(42). However, S_3 acts here on the methyl groups and is consequently the outer group of the wreath product, namely $S_3[\mathcal{E}_1 \oplus S_3]$. Finally, for $\mathcal{A}(B(CH_3)_3)$ one obtains:

$$\mathcal{A}(B(CH_3)_3) = \mathcal{E}_1 \oplus S_3[\mathcal{E}_1 \oplus S_3]. \tag{43}$$

Using this result and Tables 5 and 6, it follows that $g(\mathcal{A}) = 1 + 3(1+3) = 13$, $h(\mathcal{A}) = 1 \cdot 6(1 \cdot 6)^3 = 6^4 = 1296$.

In the case of the hypothetical dimethylboryl ammonium $(CH_3)_2BNH_3^+$, which is isotopological to trimethylborane, the permutation group given in eq.(43) is modifed to:

$$\mathcal{E}_1 \oplus S_2[\mathcal{E}_1 \oplus S_3] \oplus S_1[\mathcal{E}_1 \oplus S_3]. \tag{44}$$

The degree and the order of this automorphism group are respectively $g = 1 + 2(1+3) + 1(1+3) = 13$ and $h = 1 \cdot 2(1 \cdot 6)^2 \cdot 1(1 \cdot 6) = 2 \cdot 6^3 = 432$. The automorphism groups of isotopological molecules have equal degrees though their orders may be different.

The point groups of highest order, characterizing particular nuclear configurations of trimethylborane, are C_{3h} and C_{3v}. Both are of order 6 which is rather low compared to $h(\mathcal{A}) = 1296$. Generally, the order of the automorphism group is higher than that of the point group, provided many local symmetries occur in the molecule which are not accounted for by the point groups. In the case of high space symmetry the opposite situation is found. For the molecule BF_3 the order of the point group $h(D_{3h}) = 12$, whereas the order of the automorphism group $h(\mathcal{A}) = 6$, because definite pairs of elements of D_{3h} correspond uniquely to single permutations of $\mathcal{A}(BF_3)$. This situation becomes extreme for collinear molecules; the order of the automorphism group is 1 or 2 whereas the order of the corresponding point groups $C_{\infty v}$ and $D_{\infty h}$ is infinite.

Generally speaking, chemists regard a molecule as symmetric if the order of its corresponding point group is $h > 1$. The same idea may be intuitively applied to the symmetry properties of a graph. There exists, however, the concept of *symmetric graphs* which refers to well defined properties not generally present in graphs where $h(\mathcal{A}) > 1$. Because this concept is not pursued in current chemistry, it is not outlined here. Nevertheless, the term symmetric graph should be employed only when it refers to that concept, and not for merely indicating that $h(\mathcal{A}) > 1$.

The use of the automorphism group (based on graph theory) as a tool for describing the symmetry of nonrigid molecules represents a current frequent application of graph theory to chemistry.

Pólya's theory on the enumeration of nonequivalent isomers of a given class of compounds employs the automorphism group of a wreath of edges comprising definite vertices.

2.9 Matrix Representation and Eigenvalue Problems of Undirected Graphs

In the preceding sections graphs were represented by diagrams; here we shall employ definite matrices instead of diagrams.   The matrix representation of graphs is of particular importance in problems solved using computer assistance.   To achieve maximal clarity, this section will refer exclusively to undirected connected graphs, i.e. simple graphs.

The most fundamental properties of a graph, in addition to its number of vertices, are the pair relations stored in the edge set.   Thus $a = (1,2) \in \mathcal{G}_{28}$ means that edge a exists in $\mathcal{G}_{28}$ and is incident with vertices 1 and 2 which are therefore adjacent. All this information can also be stored in the *incidence matrix*, I.   The rows and columns of I are assigned to the vertices $v_1, v_2, \ldots, v_n$ and edges $e_1, e_2, \ldots, e_m$ of $\mathcal{G}$, respectively. The elements of I assume the value 0 and 1, according to the scheme:

$$(\mathbf{I})_{v_j e_k} = I_{jk} = \begin{cases} 1 & \ldots \text{ if } v_j \text{ is incident with } e_k; \\ 0 & \ldots \text{ otherwise .} \end{cases} \tag{45}$$

The sum of all the elements in a row indicates the degree of the corresponding vertex.   Since any edge terminates on two vertices, the sum of all the elements in a column is two. These results are expressed thus:

$$\sum_{\{k\}} I_{jk} = g(v_j); \quad \sum_{\{j\}} I_{jk} = 2. \tag{46}$$

A rearrangement of rows and columns of I corresponds to a new labelling of the vertices and edges of $\mathcal{G}$, but modifies the graph $\mathcal{G}$ only isomorphically.

For $\mathcal{G}_{28}$, (see p.59), the following incidence matrix can be derived:

$$I(\mathcal{G}_{28}) = \begin{bmatrix} 1 & 0 & 0 & 0 & 0 & 0 & 0 & 0 & 0 \\ 1 & 1 & 1 & 0 & 0 & 0 & 0 & 0 & 0 \\ 0 & 1 & 0 & 1 & 1 & 0 & 0 & 0 & 0 \\ 0 & 0 & 1 & 1 & 0 & 1 & 0 & 0 & 0 \\ 0 & 0 & 0 & 0 & 1 & 1 & 1 & 1 & 0 \\ 0 & 0 & 0 & 0 & 0 & 0 & 1 & 0 & 1 \\ 0 & 0 & 0 & 0 & 0 & 0 & 0 & 1 & 1 \end{bmatrix} \begin{matrix} 1 \\ 2 \\ 3 \\ 4 \\ 5 \\ 6 \\ 7 \end{matrix}$$
$$\phantom{I(\mathcal{G}_{28}) = } a \quad b \quad c \quad d \quad e \quad f \quad g \quad h \quad i \qquad j$$

The *degree matrix* $G(\mathcal{G})$ of a graph is a diagonal matrix of order $n(\mathcal{G})$; its diagonal elements indicate the degree of the corresponding vertices thus:

$$(G)_{v_j v_j} = G_{jj} = g(v_j). \tag{47}$$

From I and G, a further important matrix of a graph may be derived known as the *adjacency matrix*, $A(\mathcal{G})$:

$$A = I\,I^t - G. \tag{48}$$

Here, I^t designates the transpose of I. All matrices used hereafter refer to a graph with a definite labeling of the vertices and edges. From eq.(48) it follows that $A(\mathcal{G})$ is a symmetrical square matrix of order $n(\mathcal{G})$; each row and column is assigned to a vertex $v_j \in \mathcal{G}$. Moreover, from eq.(48) it follows that:

$$A_{jj} = 0;$$
$$A_{jk} = A_{kj} = \begin{cases} 1 & \dots \text{ if } v_j \text{ and } v_k \text{ are neighbors;} \\ 0 & \dots \text{ otherwise.} \end{cases} \tag{49}$$

Using these results the adjacency matrix is easily set up by inspection of the graph.

For instance,

$$A(\mathcal{G}_{28}) = \begin{bmatrix} 0 & 1 & 0 & 0 & 0 & 0 & 0 \\ 1 & 0 & 1 & 1 & 0 & 0 & 0 \\ 0 & 1 & 0 & 1 & 1 & 0 & 0 \\ 0 & 1 & 1 & 0 & 1 & 0 & 0 \\ 0 & 0 & 1 & 1 & 0 & 1 & 1 \\ 0 & 0 & 0 & 0 & 1 & 0 & 1 \\ 0 & 0 & 0 & 0 & 1 & 1 & 0 \end{bmatrix}.$$

The nonzero elements of **A** correspond to the edges of $\mathcal{G}$. Any off-diagonal element A_{jk} can be characterized by a definite direction $j \to k$ which is determined by the ordering of its indices. For the transposed element A_{kj} the reverse direction must be attributed. Consequently, an edge of a simple graph can be considered as a superposition of two oppositely directed arcs, as already mentioned in Section 1.

The adjacency matrix of a complete graph is a symmetric matrix of order n whose off-diagonal elements are all one.

Any edge (a, b) may be considered as a *walk* v_{ab} of length one. Since in a simple graph multiple edges are by definition absent, either a unique walk of length one or no walk of this length will exist between two arbitrarily selected vertices. The matrix elements of **A** precisely reflect this situation. The elements of $\mathbf{A}^2$ indicate how many walks of length two exist between the vertices considered. Generally, the value of an element of $\mathbf{A}^\nu$ is equal to the number of walks of length ν occurring between the vertices considered.

Any edge of a walk v_{ab} may be traversed several times, whereas an edge of a path w_{ab} can be traversed only once. Hence, the length of a walk between two vertices cannot be shorter than the length of the path between these two vertices. The minimal length of a path between two vertices, a and b, and the length of any walk between these vertices are therefore related as follows:

$$|w_{ab}^{\min}| \leq |v_{ab}|. \tag{50}$$

Based on this equation, a computer-assisted procedure has been developed for determining the lengths of the shortest paths in a graph. These lengths are identical to the smallest value of the exponent ν of $\mathbf{A}^\nu$ for which the corresponding matrix element $(\mathbf{A}^\nu)_{ab}$ takes nonzero values.

$$|w_{ab}^{\min}| = \min\{\nu | (\mathbf{A}^\nu)_{ab} \neq 0\}. \tag{51}$$

A very important application of eq.(51) is given in Section 11.

It can be shown generally that in the set of the powers of the adjacency matrix $\{\mathbf{A}^\nu | \nu = 0, 1, \ldots\}$ only $n = n(\mathcal{G})$ matrices are linearly independent, so that one can derive the relationship:

$$\sum_{\nu=0}^{n} a_\nu \mathbf{A}^{n-\nu} = 0. \tag{52}$$

The coefficients a_ν are identical to those of the characteristic polynomial given below in eq.(56).

By analogy with eq.(48), the following expression for the adjacency matrix of the line graph of $\mathcal{G}$ can be derived:

$$A(\mathcal{L}(\mathcal{G})) = \mathbf{I}^t \, \mathbf{I} - 2 \cdot \mathbf{1}_m \qquad (53)$$

where $\mathbf{1}_m$ represents a unit matrix of order m and $m = m(\mathcal{G})$ is the number of edges in $\mathcal{G}$.

Characteristic of a graph and of major importance are the *eigenvalues*, λ_j, of its adjacency matrix. The *eigenvalue problem* associated with a square matrix of order n is that of finding n-dimensional vectors, $\vec{c}_j$, which remain unchanged (apart from a constant factor) when multiplied by the matrix. The vectors thus have to satisfy the following equation:

$$\mathbf{A} \cdot \vec{c}_j = \lambda_j \cdot \vec{c}_j. \qquad (54)$$

Vectors satisfying this condition are called *eigenvectors*.

It can be shown that eq.(54) holds only for those values of λ for which the determinant of the square matrix $(\mathbf{A} - \lambda \cdot \mathbf{1}_n)$ is zero, i.e.

$$\det(\mathbf{A} - \lambda \cdot \mathbf{1}_n) = 0. \qquad (55)$$

Square matrices in which the elements A_{jk} and A_{kj} are either equal and real, or form a pair of complex conjugate numbers, $A_{jk} = A_{kj}^*$, are called *Hermitian matrices*. For such matrices all the eigenvalues are real numbers. In general, the adjacency matrix is a real matrix; all elements of $\mathbf{A}$ are real. However, the adjacency matrices of the representative graphs of polymers are nonreal but Hermitian (see Section 14).

In computer-assisted numerical work, the λ_j are determined by diagonalizing $\mathbf{A}$. If analytical methods are applied, the determinant given in eq.(55) is developed as a function, $\chi(\mathcal{G}, \lambda)$, of the variable λ. This function is known as the *characteristic polynomial* of $\mathcal{G}$:

$$\chi(\mathcal{G}, \lambda) = \det(\mathbf{A} - \lambda \cdot \mathbf{1}_n) = (-1)^n \sum_{\nu=0} a_\nu \cdot \lambda^{n-\nu}. \qquad (56)$$

As mentioned previously, the coefficients a_ν are identical with those of eq.(52). The eigenvalues λ_j are derived from zero values of the polynomial.

Let $G = \|G_{jk}\|$ be the determinant of eq.(56); its elements are $G_{jj} = -\lambda$ and $G_{jk} = A_{jk}$. In order to represent the expansion of the determinant concisely, the following notation will be used:

$$G = \sum_{\{P\}}(-1)^p\widehat{P}(G_{11}G_{22}\ldots G_{nn}). \qquad (57)$$

Here, $(G_{11}G_{22}\ldots G_{nn})$ is the product of all the diagonal elements and $\widehat{P}$ is the permutation operator which permutes the column indices but leaves the row indices fixed, so that, for instance, from the product $\ldots G_{jj}G_{kk}\ldots$ the product $\ldots G_{jk}G_{kj}\ldots$ is formed, etc. Furthermore, p denotes the number of transpositions corresponding to the permutations induced by $\widehat{P}$. Finally, $\{P\}$ designates the set of all $n!$ feasible permutations.

A term in the summation in eq.(57) will vanish if the permutation produced by $\widehat{P}$ contains an element G_{st} for which $G_{st} = 0$; such a situation arises when the vertices s and t are not adjacent. Thus, nonvanishing contributions to G are produced only by permutations which are composed of transpositions $\binom{u\,v}{v\,u}$, corresponding to edges $(u,v) \in \mathcal{G}$. These permutations can be represented as subgraphs of $\mathcal{G}$ known as *Sachs graphs*. In Sachs graphs, independent transpositions are represented by the corresponding complete graph K_2, and permutation cycles of order $r > 2$ are described by the appropriate cycle graph C_r. From eq.(57), all components of a Sachs graph must be vertex disjoint. The cycles C_r in Sachs graphs represent two contributions which are equivalent to each other. This is due to the fact that all cyclic permutations occur pairwise. Such a pair is formed for instance by $G_{ab}G_{bc}G_{cd}\ldots G_{fa}$ and $G_{af}\ldots G_{dc}G_{cb}G_{ba}$. Because all the elements G_{jk} are equal either to unity or to zero, the coefficients a_ν of the characteristic polynomials are determined by the number of components and the cardinality of the sets of the Sachs graphs which can be formed from ν vertices of $\mathcal{G}$. For further details the reader is directed to the corresponding chapter of the present book.

If the summation in eq.(57) is performed only over those permutations $\{P'\} \subset \{P\}$ produced by independent transpositions, the *acyclic* or *matching polynomial* $\alpha(\mathcal{G},\lambda)$ of the graph $\mathcal{G}$ is derived. The transpositions involved here have the cycle structure $[1]^{n-2r}[2]^r$. The acyclic polynomial is more extensively discussed elsewhere. It becomes important when graph theory is applied to statistics e.g. for calculation of the entropy of adsorption or for treating the Ising model.

Both polynomials can be unified to yield $\mu(\mathcal{G}, \lambda; t)$, provided the cyclic contributions in $X(\mathcal{G}, \lambda)$ are multiplied by t. For $t = 0$ and $t = 1$ the following equations hold:

$$\mu(\mathcal{G}, \lambda; t = 0) = \alpha(\mathcal{G}, \lambda);$$
$$\mu(\mathcal{G}, \lambda; t = 1) = X(\mathcal{G}, \lambda).$$

Let $\mathcal{G}$ be a simple graph, $(a, b) \in \mathcal{G}$ an edge of $\mathcal{G}$ and $\{Z_{ab}\}$ the set of all cycles in $\mathcal{G}$ comprising the edge (a, b); $(a, b) \in Z_{ab} \in \mathcal{G}$. From eq.(57) it follows that:

$$\mu(\mathcal{G}, \lambda; t) = \mu(\mathcal{G} - (a, b), \lambda; t) - \mu(\mathcal{G} - a - b, \lambda; t) - 2t \sum_{\{Z_{ab}\}} \mu(\mathcal{G} - Z_{ab}, \lambda; t)$$

$$(57')$$

where $\mathcal{G} - (a, b)$, $\mathcal{G} - a - b$, $\mathcal{G} - Z_{ab}$ are the graphs obtained from $\mathcal{G}$ by removing respectively the edge (a, b), the vertices a and b or the cycle Z_{ab}.

Generally, it can be shown that all the eigenvalues, λ_j, of $\mathbf{A}(\mathcal{G})$ are located in the range:

$$-g^{max}(\mathcal{G}) \leq \lambda_j \leq +g^{max}(\mathcal{G}), \tag{58}$$

where $g^{max}(\mathcal{G})$ is the highest vertex degree occurring in $\mathcal{G}$. If $\mathcal{G}$ is a line graph, then its eigenvalues lie in the range:

$$-2 \leq \lambda_j \leq +g^{max}(\mathcal{G}). \tag{58'}$$

Eigenvalues of graphs are of major importance in specific quantum chemical methods (see Section 13).

2.10 The Matrix Representation of Digraphs

The elements of $\mathcal{E}(\mathcal{D})$ are ordered pairs denoted by $\vec{e}_{jk} = \langle j, k \rangle \in \mathcal{D}$ which are represented in the diagrams of a digraph $\mathcal{D}$ by directed arcs which point from vertex j to vertex k. This ordering leads to some peculiarities in the matrix representation of a digraph $\mathcal{D}$ which are now elaborated.

For a digraph, the elements of the *incidence matrix* $\mathbf{I}(\mathcal{D})$ are as follows:

$$I_{v_i \vec{e}_{jk}} = \begin{cases} +1 & \dots\ i = j, \\ -1 & \dots\ i = k; \\ 0 & \dots\ \text{otherwise.} \end{cases} \tag{59}$$

This matrix is of minor importance only because, in contrast to eq.(48),

$$I\,I^t = G^+ + G^- - A - A^t,$$

where G^+ and G^- represent the matrices of the in- and outdegree, respectively. A designates the adjacency matrix of the digraph.

The elements of the adjacency matrix are defined as follows

$$A_{jk} = \begin{cases} 1 & \dots \text{ if } \langle j, k \rangle \in \mathcal{D}; \\ 0 & \dots \text{ otherwise.} \end{cases} \tag{60}$$

This equation implies that $A(\mathcal{D})$ is in general nonsymmetrical. As an example, for the digraph $\mathcal{D}_1$, the following adjacency matrix may be derived:

$$\mathcal{D}_1 : \qquad A(\mathcal{D}_1) = \begin{bmatrix} 0 & 0 & 0 & 0 \\ 1 & 0 & 1 & 1 \\ 1 & 0 & 0 & 0 \\ 0 & 0 & 1 & 0 \end{bmatrix}.$$

The sum of all the row and column elements is identical with the out- or indegree, respectively. The absolute value of the elements of the power matrix A^ν indicates the number of directed walks of length ν in $\mathcal{D}$. Since in $\mathcal{D}_1$ no directed walk of length $\nu > 3$ exists, $(A(\mathcal{D}_1))^4 = 0$.

In the analysis of digraphs the concept of *spanning sink- and source-trees*, already defined in Section 3, is of major importance. In spanning sink- and source-trees all walks originate or terminate, respectively, in a single vertex. Examples are the graphs $\mathcal{G}_7$ and $\mathcal{G}_8$. The number of spanning sink- and source-trees can be determined by means of the matrices defined as follows:

$$M^+ = G^+ - A, \qquad M^- = G^- - A. \tag{61}$$

which for $\mathcal{D}_1$ will have the form:

$$M^+(\mathcal{D}_1) = \begin{bmatrix} 0 & 0 & 0 & 0 \\ -1 & 3 & -1 & -1 \\ -1 & 0 & 1 & 0 \\ 0 & 0 & -1 & 1 \end{bmatrix},$$

$$M^-(\mathcal{D}_1) = \begin{bmatrix} 2 & 0 & 0 & 0 \\ -1 & 0 & -1 & -1 \\ -1 & 0 & 2 & 0 \\ 0 & 0 & -1 & 1 \end{bmatrix}.$$

The number of spanning sink-trees (source-trees) having the vertex v_j as sink (source) is equal to the value of the minor of the elements of the row of M^+ (column of M^-) corresponding to the vertex v_j. From the matrices above, D_1 contains only sink-trees with vertex 1 as its sink, and source-trees with vertex 2 as its source. The number of spanning sink- and source-trees is 3 and 4, respectively. They are represented below.

a) sink trees of D_1:

b) source trees of D_1:

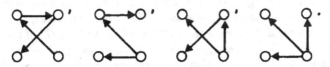

A digraph is called *Eulerian* if it contains at least one Eulerian trail. Such an Eulerian trail is defined as a spanning sequence of arcs, where any arc of D occurs only once (see Section 2). For Eulerian digraphs it is always true that:

$$M^+(D) = M^-(D) \tag{62}$$

and the sum of all the elements in any row or column is zero. From the definitions of the matrices M^+ and M^-, it can be seen that eq.(62) requires $G^+(D) = G^-(D)$, i.e. the in- and outdegrees of any vertex are identical.

2.11 Distances in Graphs and Digraphs

The concept of the length of a path or cycle has already been introduced in Section 2. It was applied to undirected and directed walks in the last two sections. In each case, the length indicates how many edges (arcs) belong to a path, a cycle or a walk. In eq.(51) a procedure is outlined for the determination of the length of shortest path.

The shortest path between any two vertices (say x and y) is called a *geodesic* and its length is described as the *distance*, $d(x, y)$, between the two vertices.

The distances occurring in a simple graph form a *discrete metric* which is defined on the vertex set of the graph. Such a metric arises because for any vertex pair $x, y \in \mathcal{G}$ a real number $d(x, y) \geq 0$ can be assigned satisfying the axioms:

[M 1] $d(x, y) = 0$, if and only if $x = y$;

[M 2] $d(y, x) = d(x, y)$;

[M 3]   *triangle inequality*: $d(x, z) \leq d(x, y) + d(y, z)$.

The metric is discrete because all distances have either integer or zero values.

It is important to mention that a metric can be imposed only upon simple connected graphs, even though distances can also be defined in digraphs. If a graph comprises more than one component, $\mathcal{G} = \mathcal{G}' \cup \mathcal{G}''$, there are no walks between the vertices $u \in \mathcal{G}'$ and $v \in \mathcal{G}''$ of the different components. Consequently, no distance can be assigned to such a vertex pair $u, v \in \mathcal{G}$. A metric is also undefined for pseudographs and digraphs. In the first instance axiom [M 1], in the second axiom [M 2] would be violated. Using particular conventions, distances can be defined even for such graphs as outlined below. The distances, however, do not form a metric.

The *distance matrix* $D(\mathcal{G})$ of a graph $\mathcal{G}$ is a square matrix of order $n = n(\mathcal{G})$; its rows and columns correspond to the vertices of $\mathcal{G}$. The elements of the distance matrix are given by the distance between the corresponding vertex pair:

$$D_{jk} = d(j, k). \tag{63}$$

From axiom [M 2], $D(\mathcal{G})$ is seen to be symmetric, whereas axiom [M 1] implies that $D_{jj} = 0$ for all j.

As an illustration, the distance matrix of the skeleton graph $\mathcal{G}_{30}$ for naphthalene (repeated here as $\mathcal{G}_{36}$) is given below:

$\mathcal{G}_{36}$:

$$D(\mathcal{G}_{36}) = \begin{bmatrix} 0 & 1 & 2 & 3 & 2 & 3 & 4 & 3 & 2 & 1 \\ 1 & 0 & 1 & 2 & 3 & 4 & 5 & 4 & 3 & 2 \\ 2 & 1 & 0 & 1 & 2 & 3 & 4 & 5 & 4 & 3 \\ 3 & 2 & 1 & 0 & 1 & 2 & 3 & 4 & 3 & 2 \\ 2 & 3 & 2 & 1 & 0 & 1 & 2 & 3 & 2 & 1 \\ 3 & 4 & 3 & 2 & 1 & 0 & 1 & 2 & 3 & 2 \\ 4 & 5 & 4 & 3 & 2 & 1 & 0 & 1 & 2 & 3 \\ 3 & 4 & 5 & 4 & 3 & 2 & 1 & 0 & 1 & 2 \\ 2 & 3 & 4 & 3 & 2 & 3 & 2 & 1 & 0 & 1 \\ 1 & 2 & 3 & 2 & 1 & 2 & 3 & 2 & 1 & 0 \end{bmatrix} \begin{array}{c} 21 \\ 25 \\ 25 \\ 21 \\ 17 \\ 21 \\ 25 \\ 25 \\ 21 \\ 17 \end{array}$$

The row sums of the distance matrix are called the *distance numbers* $D(v_j)$ of the vertex v_j. They represent the sum of the distances of one vertex v_j to all other vertices of the graph $\mathcal{G}$. For distance numbers between adjacent vertices, with $x, y \in \mathcal{G}$ and $d(x,y) = 1$, the following theorem holds:

$$D(y) = D(x) + |\mathcal{V}(x)| - |\mathcal{V}(y)|. \tag{64}$$

Here, $\mathcal{V}(x)$ and $\mathcal{V}(y)$ represent the following disjoint vertex subsets:

$$\mathcal{V}(x) = \{z | z \in \mathcal{G}, \ d(z,x) < d(z,y)\},$$
$$\mathcal{V}(y) = \{z | z \in \mathcal{G}, \ d(z,x) > d(z,y)\};$$

$\mathcal{V}(x)$ comprises those vertices z whose distance from x is smaller than that from y, whereas for vertices of $\mathcal{V}(y)$ the converse is true.

The vertex sets $\mathcal{V}(x)$ and $\mathcal{V}(y)$ are obviously disjoint subsets of $\mathcal{V}(\mathcal{G})$. In addition to these two subsets, a third disjoint subset, $\mathcal{V}(q)$, can arise, defined as follows:

$$\mathcal{V}(q) = \{z | z \in \mathcal{G}, \ d(z,x) = d(z,y)\}.$$

This subset collects all the vertices of $\mathcal{G}$ having equal distances from x and y. The subset $\mathcal{V}(q)$ is nonempty if and only if the edge (x,y) belongs to an odd membered cycle of $\mathcal{G}$. The distance numbers of $\mathcal{G}_{36}$ are recorded in Figure 2; they may be used in order to verify the theorem given by eq.(64).

A *distance tree* $T(\mathcal{G}, v_j)$ of the vertex $v_j \in \mathcal{G}$ is defined as a tree spanning $\mathcal{G}$ in which the distance number of the vertex v_j is the same as in $\mathcal{G}$. In Figure 2 three of the 40 distance trees of $\mathcal{G}_{36}$, some of which are isomorphic, are represented.

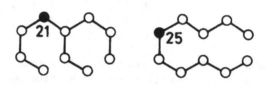

Distance Numbers and Some of the Distance Trees for the Skeleton Graph of Naphthalene, $\mathcal{G}_{36}$.

The sum of all the elements in a triangle of the distance matrix is called the *Wiener number*, $W(\mathcal{G})$, of the graph, i.e.

$$W = \sum_{j<k} \sum d(j,k) \qquad (65)$$

$W(\mathcal{G})$ was the first topological index introduced into chemistry. Obviously, the sum of the distance numbers of all vertices equals exactly $2W$. The distance numbers are a useful tool for investigating the relationships between the Wiener number and the topology of a graph.

In order to formulate a distance matrix for unconnected graphs and digraphs, the distance between two vertices which are not connected by a walk is defined as ∞ (in infinite simple graphs geodesics of infinite length may also occur). Hence, the distance matrix of a graph composed of two components, $\mathcal{G} = \mathcal{G}' \cup \mathcal{G}''$, will exhibit the following block form:

$$D(\mathcal{G}' \cup \mathcal{G}'') = \begin{bmatrix} D(\mathcal{G}') & \infty \\ \infty & D(\mathcal{G}'') \end{bmatrix},$$

where ∞ denotes a matrix with the elements ∞.

The elements of the distance matrix of a digraph $\mathcal{D}$ are defined as:

$$D_{zz} = 0, \quad D_{xy} \doteq \begin{cases} |\vec{w}_{xy}^{\,min}| & \dots \text{ if } \vec{w}_{xy}^{\,min} \in \mathcal{D} \\ \infty & \dots \text{ otherwise.} \end{cases}$$

Here $\vec{w}_{xy}$ represents a directed path which originates at x and terminates at y. For the numerical determination of $|w_{xy}^{min}|$ eq.(51) can be applied.

The elements of the *detour matrix*, $U(\mathcal{D})$, are defined by the length of the longest directed path, $|\vec{w}^{max}|$. If for a vertex pair $x,y \in \mathcal{D}$, the element $D_{xy} = \infty$, then $U_{xy} = \infty$ also holds.

An additional matrix used for digraphs is the *reachability matrix*, $R(\mathcal{D})$, whose elements assume the following form:

$$R_{xx} = 1, \; R_{xy} = \begin{cases} 1 & \dots \text{ if } \vec{w}_{xy} \in \mathcal{D} \\ 0 & \dots \text{ otherwise.} \end{cases}$$

The information stored in $D(\mathcal{D})$ is contained in $R(\mathcal{D})$ in a simplified manner: For all vertex pairs $x,y \in \mathcal{D}$ corresponding to $D_{xy} = \infty$, $R_{xy} = 0$; all other elements of R are equal to 1.

In order to illustrate the above extension, the three matrices are presented below for the digraph $\mathcal{D}_1$ illustrated in the previous section.

$$D(\mathcal{D}_1) = \begin{bmatrix} 0 & \infty & \infty & \infty \\ 1 & 0 & 1 & 1 \\ 1 & \infty & 0 & \infty \\ 2 & \infty & 1 & 0 \end{bmatrix},$$

$$U(\mathcal{D}_1) = \begin{bmatrix} 0 & \infty & \infty & \infty \\ 3 & 0 & 2 & 1 \\ 1 & \infty & 0 & \infty \\ 2 & \infty & 1 & 0 \end{bmatrix},$$

$$R(\mathcal{D}_1) = \begin{bmatrix} 1 & 0 & 0 & 0 \\ 1 & 1 & 1 & 1 \\ 1 & 0 & 1 & 0 \\ 1 & 0 & 1 & 1 \end{bmatrix}.$$

2.12 Metric and Topological Spaces for Simple Graphs

In the previous section we discussed the notion of a discrete metric imposed upon the vertex set of a simple graph. Because a set and its imposed metric form a metric space, *any simple graph is associated uniquely with a metric space* $\mathcal{R}(\mathcal{G})$. The vertex set of G is the basis of the metric space $\mathcal{R}(\mathcal{G})$. The vertices are called "the points of the space". The edge set $\mathcal{E}(\mathcal{G})$ of the graph determines the structure of $\mathcal{R}(\mathcal{G})$.

The *ball neighborhood* $\mathcal{U}_\varepsilon(p)$ of a point (vertex) p is the subset of the vertex set $\mathcal{V}(\mathcal{G})$ defined as follows:

$$\mathcal{U}_\varepsilon(p) = \{x | x \in \mathcal{G}, \, d(p,x) < \varepsilon\}, \tag{66}$$

where $\varepsilon > 0$ represents an arbitrary real number. An immediate consequence of this definition is that point p belongs to each of its ball neighborhoods, $p \in \mathcal{U}_\varepsilon(p)$ for all ε because $d(p,p) = 0 < \varepsilon$. Because of the discrete metric only integral values are meaningful for ε. For the vertex $p = 1$ of graph $\mathcal{G}_{36}$ (Section 11) the following ball neighborhoods can be derived:

$$\mathcal{U}_1(1) = \{1\},$$
$$\mathcal{U}_2(1) = \mathcal{U}_1(1) \cup \{2, 10\},$$
$$\mathcal{U}_3(1) = \mathcal{U}_2(1) \cup \{3, 5, 9\},$$
$$\mathcal{U}_4(1) = \mathcal{U}_3(1) \cup \{4, 6, 8\},$$
$$\mathcal{U}_5(1) = \mathcal{U}_4(1) \cup \{7\} = \mathcal{V}(\mathcal{G}).$$

A subset $\mathcal{U} \subset \mathcal{V}(\mathcal{G})$ is called a neighborhood of p if it comprises a ball neighborhood of p : $\mathcal{U}_\varepsilon(p) \subset \mathcal{U}$. Thus, in case of a metric space $\mathcal{R}(\mathcal{G})$ associated with the simple graph $\mathcal{G}$ any subset $\mathcal{U} \subset \mathcal{V}(\mathcal{G})$ containing p is a neighborhood of p.

The following important four conditions are satisfied by neighborhoods:

(U 1) p is contained in each neighborhood $\mathcal{U}$ of p;

(U 2) if $\mathcal{U}$ is a neighborhood of p, any superset $\mathcal{U}' \supset \mathcal{U}$ is also a neighborhood of p;

(U 3) the intersection $\mathcal{U}' \cap \mathcal{U}''$ of two neighborhoods of p is also a neighborhood of p;

(U 4) a neighborhood $\mathcal{U}$ of p is also a neighborhood of all points of an appropriately selected neighborhood $\mathcal{U}'$ of p.

The first three conditions arise from the definition of a neighborhood; (U 4), however, requires more elaboration. Obviously, if $\mathcal{U}_\varepsilon(p) \subset \mathcal{U}$ and $x \in \mathcal{U}_\varepsilon(p)$, a ball neighborhood $\mathcal{U}_\eta(x)$ exists which is contained in $\mathcal{U}_\varepsilon(p)$.

For a vertex $p \in \mathcal{G}$ a system of neighborhoods, $\underline{U}(p)$, can be constructed which satisfy the conditions (U 1) to (U 4). For any vertex of $\mathcal{G}$ the construction of such a system of neighborhoods is possible. This fact has an important consequence, namely that *any simple graph $\mathcal{G}$ is also associated with a topological structure, $T(\mathcal{G})$.* A topological structure is imposed on a set by forming for each member p of the set the system of neighborhoods, $\underline{U}(p)$, where the neighborhoods are subsets of the original set satisfying the axioms below:

[U 1] $p \in \mathcal{U}$ for any $\mathcal{U} \in \underline{U}(p)$;

[U 2]   if $\mathcal{U} \in \underline{U}(p)$ and $\mathcal{U}' \supset \mathcal{U}$, then also $\mathcal{U}' \in \underline{U}(p)$;

[U 3]   if $\mathcal{U}', \mathcal{U}'' \in \underline{U}(p)$, then also $\mathcal{U}' \cap \mathcal{U}'' \in \underline{U}(p)$; $\mathcal{V}(\mathcal{G}) \in \underline{U}(p)$;

[U 4]   for $\mathcal{U} \in \underline{U}(p)$, an $\mathcal{U}' \in \underline{U}(p)$ exists, such that $\mathcal{U} \in \underline{U}(x)$ holds for all $x \in \mathcal{U}'$.

Since a set together with its topological structure is called a topological space, any simple graph will be associated with such a topological space $T(\mathcal{G})$. The use of the same symbol for topological structure and for topological space should not cause any confusion. Consequently, one calls $\mathcal{R}(\mathcal{G})$ a topological space with a metric, and conversely one says the topological structure $T(\mathcal{G})$ is induced by the metric of $\mathcal{G}$.

Axioms [U 1] to [U 4] were introduced by *Hausdorff*. The topologies defined by these axioms are called *neighborhood topologies*. Recently, another procedure for the association of a topological space with a simple graph, using an open set formalism for defining topological structures, was reported by *Merrifield and Simmons*.

At the end of this section the question is posed: What impact on chemistry has the fact that any simple graph, in particular a constitutional graph, is associated with a topological space? Currently, this issue is more often raised than resolved. The constitutional formulas of chemical compounds are represented by simple graphs. Consequently, for any chemical compound a unique topological space can be assigned. In the reverse direction, however, this assignment is not unique. Graph $\mathcal{G}_1$ introduced at the beginning of this chapter is the constitutional graph of isopentane, dimethylethylchlorosilane, and other compounds having the same topology. Each of these compounds will belong to a different class of compounds, viz hydrocarbons, alkyl-substituted monosilanes, etc., respectively. All molecules having isomorphic constitutional graphs have the same topology and are called *isotopological*. On the one hand, the constitutional graph corresponds to a set of isotopological compounds, $C(\mathcal{G}) = \{C_1, C_2, \ldots\}$, and on the other it is assigned a single topological space $T(\mathcal{G})$ in which the properties of those compounds can be studied. Provided a topological property $\underline{T}$ can be quantified, the derived number T will be valid for all elements of $C(\mathcal{G})$. Thus, T is valid for any *isotopological compound* corresponding to G, regardless of the chemical

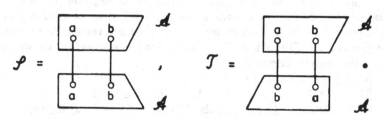

$$\mathcal{S} = \qquad , \qquad \mathcal{T} =$$

The Most Simple TEMO Model

nature of the individual compounds.

If for any class of compounds a measurable physical quantity P can be represented by a function $P = P(T)$, for instance

$$P = AT + B,$$

where T is such a topological quantity (topological index), this function must be valid for all classes of compounds if the correlation is of topological origin. Only the constants A and B may vary for the various classes according to the chemical nature of the compounds forming the class. To the best of our knowledge, investigations along these lines have not been pursued in the current chemical literature.

Recently a *topological effect on molecular orbitals* (TEMO) has been discovered. It interrelates the eigenvalues of two graphs, S and T, having a definite topological relationship, namely, that their topological spaces, $T(S)$ and $T(T)$, may be partitioned into two or three subspaces as follows:

$$T(S) = T(S_1) \oplus T(S_2) \oplus \dots$$
$$T(T) = T(T_1) \oplus T(T_2) \oplus \dots$$

such that the subspaces of S and T are pairwise isomorphic, i.e.

$$T(S_j) \cong T(T_j).$$

Thus, the difference of these two topological spaces, $T(S)$ and $T(T)$, originates only from the connections of their pairwise isomorphic subspaces. For simplicity, in the example below we assume $T(S_1) \cong T(S_2) \cong T(T_1) \cong T(T_2)$ and that the graphs corresponding to these subspaces are uniquely denoted by A. The connection of two subunits of S and T, respectively, is performed by two edges which are incident to nonequivalent vertices, $a, b \in A$: In S the edges (a, a) and (b, b), and in T the edges (a, b) and (b, a) are formed as shown in Figure 3. For instance, *o-* and *p*-benzoquinodimethane represent such a pair of S and T isomers.

The eigenvalues $\{\lambda_j^S\}$ of S and $\{\lambda_j^T\}$ of T satisfy the following inequalities:

$$\lambda_1^S \leq \lambda_1^T \leq \lambda_2^T \leq \lambda_2^S \leq \lambda_3^S \leq \lambda_3^T \leq \lambda_4^T \leq \lambda_4^S \leq \dots . \qquad (67)$$

In the intervals defined by two consecutive eigenvalues of S there are alternately two or no eigenvalues of T. Eq.(67) represents a novel interlacing theorem which is not restricted to the eigenvalue spectra of such graphs. Because for completely conjugated unsaturated systems the eigenvalues λ_j^S and λ_j^T may be identified with π-MO energies, the interlacing should be substantiated by photoelectron spectra; this is verified by numerous examples. For more details the reader may wish to consult the papers cited at the end of this chapter.

2.13 Graphs in Quantum Chemistry

Simple graphs were introduced into quantum chemistry insofar as they are related to the Hückel HMO procedure. The adjacency matrix of a skeleton graph for a complete conjugated hydrocarbon is identical with the HMO matrix, provided the energy scale is properly selected. More correctly, however, the graph of the HMO matrix is a graph representing the π-AO basis and its interactions. Owing to its isomorphism with the skeleton graph, both graphs pertain to the same topological space.

In topological considerations of the σ-electron system, difficulties arise when the s and p_σ atomic orbitals are used as a basis. Therefore, a recently developed method for the topological treatment of σ-electrons uses localized bond orbitals as basis functions which correspond to the edges of the constitutional graph. Thus, the line graph of a constitutional graph serves in this method as a basis graph for the σ molecular orbitals.

2.14 Weighted Graphs

Sometimes the need arises to assign weights to the vertices and edges of simple graphs in order to simulate real situations. Thus, in HMO calculations particular vertices mapping heteroatoms may be characterized by a weight which depends on the value of their Coulombic integrals, α_r, as used in the HMO method. The so-called "geometrical" parametrisation of HMO-calculations is performed by a weighting of the edges, β_{rs}. Another example is furnished by the HMO ω SCF technique.

Using this technique, the charge and bond densities obtained in one step of the calculation are transferred into a weighting of the vertices and edges. This weighting procedure is repeated until self-consistency of all the weights is achieved.

In *representative graphs of regular polymers* the monomer unit is depicted by a simple graph, and the bonds between adjacent monomer units are indicated by arcs which are weighted by conjugate complex numbers. The characteristic polynomial of the representative graph, and the Bloch type crystal orbital description of the polymer, are equivalent.

In all these cases the weighting of graphs can be represented by the weight matrix W. The latter differs from the adjacency matrix because its diagonal elements are not necessarily zero and the off-diagonal elements can be nonintegral numbers. The eigenvalue problem for the weight matrix W is formulated similarly to that for the adjacency matrix A in eq.(55). The expansion of the determinant as given in eq.(57) is applicable. Obviously, in this case the coefficients of the characteristic polynomial are functions of the vertex and edge weights of the graph.

2.15 Bibliography

Some titles are recommended below which should lead to a deeper understanding of graph theory, its applications, and its topological aspects. Because of the extensive reference sets given elsewhere in this monograph, the citing of original communications is avoided here. Only four papers concerning group theoretical treatments of nonrigid molecules are cited. In the case of TEMO, briefly outlined only in Section 12 of this chapter, a few references are also given.

Textbooks and Monographs on Graph Theory and Topology.

1. D. König, *Theorie der endlichen und unendlichen Graphen. Kombinatorische Topologie der Streckenkomplexe*, Chelsea Publ., New York (Original version: Leipzig: Akadem. Verl.-Ges. 1936).
2. C. Berge, *The Theory of Graphs and its Applications*, Methuen, London 1962.
3. R.G. Busacker and T. Saaty, *Finite Graphs and Networks*, McGraw-Hill, New York 1965.
4. J. Dugundji, *Topology*, Allyn and Bacon, Boston 1966.
5. W. Knödel, *Graphentheoretische Methoden und ihre Anwendungen*, Springer, Berlin, Heidelberg, New York 1969.

6. H. Sachs, *Einführung in die Theorie der endlichen Graphen*, Hanser, München 1971.
7. F. Harary, *Graph Theory*, Addison-Wesley, Reading (Mass.) 1972; F. Harary, *Graphentheorie*, Oldenbourg, München 1974.
8. N. Biggs, *Algebraic Graph Theory*, Cambridge Univ. Press, London 1974.
9. W. Rinow, *Lehrbuch der Topologie*, Dtsch. Verl. Wissenschaften, Berlin 1975.
10. A.T. Balaban (Ed.), *Chemical Applications of Graph Theory*, Academic Press, New York 1976.
11. A. Graovac, I. Gutman and N. Trinajstić, *Topological Approach to the Chemistry of Conjugated Molecules*, Springer, Berlin 1977.
12. J. Hinze (Ed.), *The Permutation Group in Physics and Chemistry*, Springer, Berlin 1979.
13. D.M. Cvetković, M. Doob and H. Sachs, *Spectra of Graphs - Theory and Applications*, Academic Press, New York 1980.
14. H.N.V. Temperley, *Graph Theory and Applications*, Ellis Horwood, Chichester 1981.
15. N. Trinajstić, *Chemical Graph Theory* (2 Vols), CRC Press, Boca Raton 1983.
16. R.B. King (Ed.), *Chemical Applications of Topology and Graph Theory*, Elsevier, Amsterdam 1983.
17. I. Gutman and O.E. Polansky, *Mathematical Concepts of Organic Chemistry*, Springer, Heidelberg 1986.

Nonrigid Molecules, etc.

1. J. Serre, "Symmetry groups of nonrigid molecules," Adv. Quantum Chemistry 8 (1974), 1.
2. K. Balasubramanian, "Recent applications of group theoretical generators to chemical physics," Croat. Chem. Acta 57 (1984), 1525.
3. D.M. Walba, R.M. Richards and R.C. Haltiwanger, "Total synthesis of the first molecular Möbius strip," J. Amer. Chem. Soc. 104 (1982), 3219.
4. R.E. Merrifield and H.E. Simmons, "Structures of molecular topological spaces," Theoret. Chim. Acta 55 (1980), 55.

Topological Effect on MO's

1. O.E. Polansky and M. Zander, J. Mol. Struct. 84 (1982), 361.
2. O.E. Polansky, M. Zander and I. Motoc, Z. Naturforsch. 38a (1983), 196.

3. W. Fabian, I. Motoc and O.E. Polansky, Z. Naturforsch. 38a (1983), 916.

4. A. Graovac, I. Gutman and O.E. Polansky, Mh. Chemie 115 (1984), 1.

5. I. Motoc, J.N. Silverman and O.E. Polansky, Phys. Rev. A28 (1983), 3673.

6. I. Motoc, J.N. Silverman and O.E. Polansky, Chem. Phys. Lett. 103 (1984), 285.

7. O.E. Polansky, J. Mol. Struct. 113 (1984), 281.

8. A. Graovac and O.E. Polansky, Croat. Chem. Acta 57 (1984), 1595.

9. I. Motoc and O.E. Polansky, Z. Naturforsch. 39b (1984), 1053.

10. I. Motoc, J.N.Silverman, O.E. Polansky and G. Olbrich, Theoret. Chim. Acta 67 (1985), 63.

11. I. Gutman, A. Graovac and O.E. Polansky, Chem. Phys. Lett. 116 (1985), 206.

12. A. Graovac, I.Gutman and O.E. Polansky, J. Chem. Soc., Faraday Trans. II, 81 (1985), 1543.

13. O.E. Polansky, Match (Math. Chem.) 18 (1985), 111, 167, 217.

14. O.E. Polansky and G. Mark, Match (Math. Chem.) 18 (1985), 249.

15. M. Zander and O.E. Polansky, Naturwiss. 71 (1984), 623.

Acknowledgment

The assistance of Mrs. I. Heuer, Mrs. R. Speckbruck and Mrs. E. Currell is gratefully acknowledged.

Chapter 3

NOMENCLATURE OF CHEMICAL COMPOUNDS

Alan L. Goodson

Chemical Abstracts Service, P.O. Box 3012, Columbus, OH 43210

3.1 Introduction

The ready communication of structural information is fundamental to the development of chemistry. The most universally understood form of such information is the chemical structure diagram, such as that illustrated in Figure 1. However, it is frequently inconvenient to convey structural information directly (e.g., in conversation), so a number of other methods of representing chemical structures have been developed to satisfy a variety of needs. These methods include nomenclatures, notations, connection tables, adjacency matrices, molecular formulas, and fragment codes [1,2].

The term "nomenclature" is sometimes used as a generic term to include not only nomenclatures but also notations and connection tables. Since reference is made in this chapter to nomenclatures and notations, it is useful to define them here and to distinguish them from connection tables (see Figure 1).

Nomenclature is defined [3a] as: "a system or set of names or designations used in a particular science, discipline, or art and formally

Chemical Structure Diagram

Nomenclature

Hydroquinone or 1,4-Benzenediol

Notation

Dyson/IUPAC Line Notation: B6Q14

Wiswesser Line Notation: QR DQ

Connection Table

```
TOPOLOGY
    ATOM NO          1   2
    CONN
    ELEMENT          O   O
    BOND
    RING CORRESPONDENCE LINK
    RING ID                  46T.150A.182
    RING ATOM NOS             1   2   3   4   5   6
    SUBSTANCE ATOM NOS       03  04  05  06  07  08

    LINK GROUP               01-1 03,  02-1 08

RING
    RING IDENTIFIER NO = 46T.150A.182
    ATOM NO          1   2   3   4   5   6
    CONN               01  01  02  03  04
    ELEMENT         C  C   C   C   C   C
    BOND               *5  *5  *5  *5  *5
    RING CLOSURE PAIRS  05*5  06
```

Figure 1. Various methods of representing a chemical structure.

adopted or sanctioned by the usage of its practitioners." Chemical nomenclature is more specifically defined as: "a set of chemical names that may be systematic ... or not and that aims to tell the composition and often the structure of a given compound by naming the elements, groups, radicals, or ions present and employing suffixes denoting function ..., prefixes denoting composition ..., configuration prefixes ..., operational prefixes ..., arabic numbers or Greek letters for indicating structure (as positions of substituents), or Roman numerals for indicating oxidation state."

Chemical nomenclature is illustrated in Figure 1 with reference to the chemical structure diagram mentioned above. The substance depicted is commonly known as hydroquinone but is also indexed as 1,4-benzenediol.

A notation is defined [3b] as: "a system of characters, symbols, or abbreviated expressions used in ... science to express technical facts, quantities or other data." To illustrate notations, the IUPAC-Dyson Line Notation [4] for hydroquinone (1,4-benzenediol) is B6Q14 and the corresponding Wiswesser Line Notation (WLN) [5] is QR DQ (see Figure 1).

A connection table has been defined [6] as "a uniquely ordered list of the node symbols of the structure (or graph) in which the value (atomic symbol) of each node and its attachment (bonding) to the other nodes of the total structure are described", though it does not have to be unique. The Chemical Abstracts Service (CAS) Registry III connection tables [7] are unique, however, and that for hydroquinone is illustrated in Figure 1.

Connection tables, notations, and nomenclatures are of value in different ways and an in-depth discussion of their use has been published recently [8]. Connection tables, being atom-by-atom computer records of chemical structures, are useful for machine registration of chemical structures and for substructure searching, as in CAS ONLINE [9,10]. Chemical line notations, such as WLN's, are also convenient tools for registration and substructure searching of files of moderate size (50,000 –100,000 structures), although they have been used for some large files (exceeding 1 million structures). Nomenclatures are of value for oral and written communication. They have wider application and greater flexibility than chemical line notations and connection tables, which can only be used to describe completely known structures, because substances can be given unambiguous trivial names even if their structures are unknown or are only partially known.

Nomenclatures can be divided into three classes, i.e., systematic, semisystematic, and nonsystematic or "trivial" (see Figure 2). Systematic nomenclature, like notations and connection tables, provides a direct translation between name and structure, trivial (nonsystematic) nomenclature does not, and semisystematic nomenclature is a mixture of the two.

Systematic nomenclature is preferable for indexing purposes because names of related structures can be grouped reasonably close together in a highly ordered, sorted index. The structures of the relatively few basic component parts can then be illustrated, providing an easy translation of a name to the corresponding chemical structure. Thus, the systematic name illustrated in Figure 2 is indexed at 1H-Cyclopenta[*a*]phenanthrene in *Chemical Abstracts (CA)*, together with a structure diagram and other

Chemical Structure Diagram

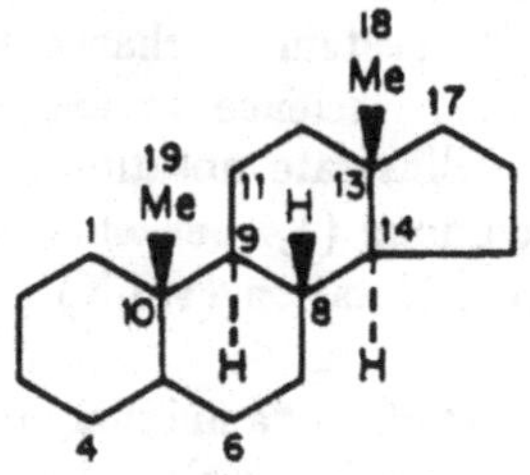

Systematic Name

* [8S-8α,9β,10α,13α,14β] — Hexadecahydro-10,13-dimethyl-
  1H-cyclopenta[a]phenanthrene

Trivial Name

Androstane

Chemical Structure Diagram

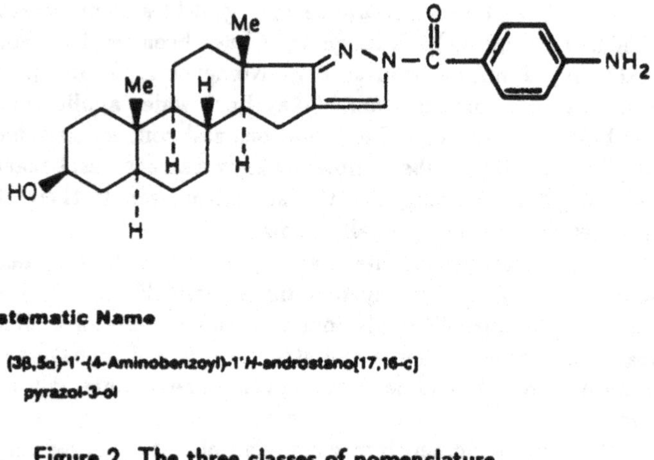

Semisystematic Name

(3β,5α)-1'-(4-Aminobenzoyl)-1'H-androstano[17,16-c]
pyrazol-3-ol

Figure 2. The three classes of nomenclature.

derivatives of that ring system.

Trivial nomenclature provides short, simple names but make considerable demands on the human memory and makes necessary a dictionary to match the trivial names with the chemical structures they represent.

Trivial names, such as "Androstane" in Figure 2, are therefore useful in small files and in indexes for representing complex structures with extensive stereochemistry. However, it would be impracticable to use only trivial names in indexes because that would require including, say, six million structure diagrams to define six million names in order to reduce the demands on the human memory.

Semisystematic nomenclature is also of use in small files and in indexes for representing, in a more compact manner, derivatives of complex structures with extensive stereochemistry, as shown in Figure 2 for a derivative of androstane.

Some argue that a chemical name should be short and easy to pronounce and that complex names are really notations. According to this view, ethanol is a true name but

2,2'-[[2,2-bis[[[2-(sulfooxy)ethyl]sulfonyl]methyl]

-1,3-propanediyl]bis(sulfonyl)]bisethanol

is a notation because the parentheses, brackets, and locants are not easily pronounceable. However, such a complex name merely reflects the complexity of the chemical structure it represents. The parentheses and brackets are necessary to avoid ambiguity and the locants are necessary to specify which isomer is being named. Further, to label one a name and the other a notation raises the question of where the distinction is to be made, any answer to which would be arbitrary.

Names are the oldest and most common means of access to the chemical literature (e.g., through *CA* indexes), although direct online access to chemical structures is now expanding rapidly. The terminology used in such indexes must reflect general usage and must be at least reasonably familiar to chemists searching the indexes. In the past, trivial names were used extensively for this reason, but more recently it has become less practical to do so and use of trivial names has declined in favor of more systematic names.

3.2 Development of Chemical Nomenclature

The origins of chemical nomenclature can be traced back to ancient China and Egypt [11–14]. Since then, chemical nomenclature has developed in a cyclical fashion, with existing practices being adequate for a time, then becoming progressively more inadequate until a crisis develops, and then the cycle being completed by introduction of a novel system that, in turn, proves adequate for a time.

It is not the function of this chapter to trace the development of chemical nomenclature in detail; this has already been done by Crosland in a very readable book [11], where references to the literature can be found. But it is useful to highlight here some important historical developments.

The first development of importance was the proposal of the first systematic nomenclature, principally for inorganic compounds, in 1787. This proposal introduced such names as sodium sulfate and potassium hydrogen tartrate, despite the lack of a supporting theory.

The first step in the development of a supporting theory was the introduction, by Dalton in 1803, of symbols representing single atoms rather than any amount of an element. This led to the first attempts to represent chemical structures by structure diagrams (Figure 3). The structure diagrams provided the needed theoretical basis for the recently-proposed systematic nomenclature and laid the foundation for the continued development of systematic nomenclature and for the eventual introduction and development of notations and connection tables.

Berzelius took the matter one step further in 1813 by proposing replacement of the alchemical-like symbols of Dalton with one- and two-letter abbreviations of element names (Figure 3), thus paving the way for modern chemical structure diagrams.

Butlerov recommended the use of structure diagrams for organic compounds in 1861. Three years later, Lothar Meyer published recognizably modern structure diagrams for water, ammonia, and methane; and Crum Brown the same for ethane. The following year (1865), Kekulé first published the hexagonal structure diagram for benzene (Figure 3).

The three-dimensional structure of molecules was postulated independently by van't Hoff and le Bel in 1874, thus giving rise to the concept of stereochemistry.

The first concerted international effort to develop coherent policies for systematic organic nomenclature was made by the International Commission for the Reform of Chemical Nomenclature at the Geneva Congress in 1892. From the Commission developed what is known today as the International Union of Pure and Applied Chemistry (IUPAC), whose declared function is to systematize and codify existing nomenclature practices insofar as they are determined to be sound. Such efforts have been well documented [15] and have resulted in the publication of comprehensive rules for systematic inorganic and organic nomenclature [16,17]. Paren-

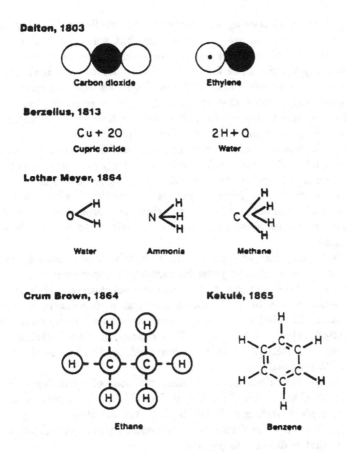

Figure 3. Development of chemical structure diagrams.

thetically, it should be noted here that, because of its need to name in a consistent manner new substances as they are reported, CAS has also been active for many years in development of chemical nomenclature [18–21], in cooperation with other organizations such as IUPAC.

It seems that chemical nomenclature is again in the position that existing nomenclature practices are becoming progressively more inadequate and that it may become necessary once again to resolve the developing crisis by introducing a novel nomenclature system, preferably one that is comprehensive. It is currently estimated that a total of about seven million substances have been recorded in the literature and the number is growing steadily. Recording and retrieving information concerning these substances requires the use of a computer, but when Elk [22] tried to translate the IUPAC rules for naming polycyclic aromatic hydrocarbons into a working algorithm, he found that there exist inherent inadequacies in some of the pertinent rules. These inadequacies limit the capabilities of the algorithm because they create ambiguous or contradictory names for various polybenzene systems. Elk introduced a novel nomenclature for polybenzene systems which he was able to translate into an algorithm. To date, CAS has not found it economically feasible to use the computer for naming new substances according to its existing practices, despite the fact that an algorithm has been written for doing so [23]. The best that CAS has achieved so far [24] is to retrieve a name that has been manually generated and recorded on file, and to extend this procedure by identifying substance names already on file that are related to a new substance being named.

It is logical that any novel, comprehensive nomenclature system would permit use of the computer for naming the large number of known substances and for naming newly-recorded substances. One step in this direction would be to formalize chemical nomenclature by use of mathematical linguistics, in which nomenclature rules are compared with grammatical rules of a language. This concept has been discussed by Mestyanek and Réti [25] and has been used by Krishnamurthy in his WISENOM proposal [26].

However, the introduction of online methods of retrieving information, such as CAS ONLINE [9,10] and DARC [27] (which provide convenient and rapid searching of large files with the aid of graphic input and output), will reduce the future need for nomenclatures and notations by an extent that is difficult to predict.

Because chemical nomenclature developed alongside chemistry, it is not surprising that existing practices are, in fact, a mixture of nomenclature systems with overlapping rules, inconsistencies, and inadequate treatment for some classes of compounds. That these problems exist and that existing practices have been becoming progressively more inadequate have been recognized for many years, resulting in a number of calls for improvement in chemical nomenclature.

Among those describing problems with existing nomenclature and discussing the need for improvements are Elk [22], Scott [28], Taylor [29], Terent'ev *et al* [30], Read and Milner [31,32], Balaban [33,34], Dyson [35], Rush [36], Cahn and Dermer [37], Lozac'h *et al* [38], Goodson [39,40], and Bonchev [41]. Efforts to resolve these problems by modification of existing practices have been discussed by Verkade [15], Loening [42], Fletcher [43], and by Fernelius *et al* [44]. An alternative to continual modification of existing practices would be to develop a new, systematic nomenclature from first principles. This approach has been discussed by Fletcher [43] and by Fernelius *et al* [44] who concluded that, despite the recognized deficiences, there is no alternative but to continue with existing nomenclature practices. However, this conclusion was reached before the full impact of the computer on chemical information storage and retrieval was felt.

Subsequently, Cahn and Dermer [37] recognized that introduction of any new nomenclature cannot be contemplated unless it is capable of manipulation by computer. The computer also provides the opportunity for any new system to be thoroughly tested before the decision is made whether or not it should be adopted by the chemical community.

Nomenclature systems based on well-defined structures such as hydrocarbons [17–21,29,30,43] prove inadequate when structures for which they have no model or parent structure are encountered. The less information the parent structures contain, the more flexible is the terminology and the more easily can it accommodate novel types of chemical structures. Ideally, any novel nomenclature system should be based on parent structures that indicate merely that two or more unspecified atoms are connected by unspecified bonds; this requirement is satisfied by chemical graphs.

3.3 Development of Chemical Line Notations

In this section and in Section 5, the concepts embodied in the notations and nomenclatures reviewed are described in the text and illustrated in the figures. The descriptions are of necessity brief and are intended for comparison purposes only: they are not intended to replace the original documents. It should also be mentioned that where names in the figures have had to be continued on the next line, the second part of the name should be regarded as closed up to the first. Classical names are included in the figures for comparison purposes where they have been supplied by the authors.

During the period 1930–5, Siboni and Perino developed a form of systematic nomenclature (Figure 4) which was reviewed by Verkade [45] (who provided a complete bibliography to which the reader is referred) and which he called a "play on letters" and an "assembling nomenclature".

Atoms and groups of a chemical structure were represented by one through four letters and these morphemes were linked together in a prescribed manner to yield the name of the structure. Locants, required for the names of the more complex structures, were placed before, after, and below the name, never within it. Thus, in the first name, *kal* means potassium, *abo* single valent, *ac* six valent, *sulf* sulfur, and *aeto* dibasic acid; in the second name, *n* means methylene, *an* methyl, and *ago* carboxamide; and in the third name, there is an ethylene group in the 2,3-position (2 *ev*), a trimethylene group (*t*), a ring (*ocl*), an aldehyde group attached directly to atom 5 (5_1 *ad*), and an oxime group attached to atom 1 (1 *afo*). While the Siboni–Perino proposals contained some features possessed by notations (in particular, linking morphemes together in a prescribed manner), it is not known whether they influenced their development.

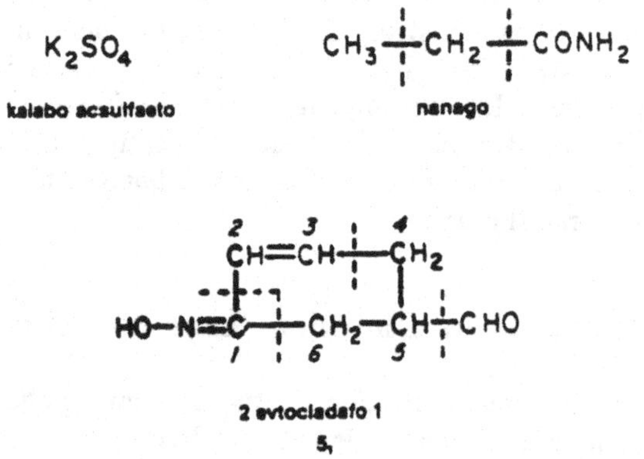

$$K_2SO_4 \qquad CH_3 \!-\! CH_2 \!-\! CONH_2$$

kalabo acsulfaeto   nanago

$$\begin{array}{c} \overset{2}{CH}=\overset{3}{CH} \!-\! \overset{4}{CH_2} \\ HO\!-\!N\!\equiv\!\overset{1}{C} \!-\! \overset{6}{CH_2}\!-\!\overset{5}{CH}\!-\!C\,HO \end{array}$$

2 evtocladafo 1
5_1

Figure 4. Siboni-Perino nomenclature.

However, we do know that notations were first suggested by Dyson in October 1946 [46–48], a result of his attention having been directed in 1937 [35] to the complexity of systematically naming organic structures. His notation was adopted by IUPAC and published in 1958 as a tentative international chemical notation [49] and in 1961 as an approved notation [50]. It was revised in 1968 to accommodate experience obtained from computer application of the notation [4] and revised again for the last time in 1976 [35] when it was expanded to include a nomenclature system.

Other notations have been proposed [51] but the only one to be used extensively has been that proposed by Wiswesser [5,52]. However, one unusual notation is worthy of mention because it is a type of linear chemical formula. Read and Milner [31,32] have developed a method for generating unique line formulas for acyclic structures that is suitable for both manual and machine encoding and decoding. Their coding algorithm is automatic and does not require chemical intuition. The unique notations generated by their algorithm, such as the first shown in Figure 5, are readily recognizable by chemists. Read has extended the concept to ring systems with substituents [53]. For example, the second notation shown in Figure 5 describes a structure that can be constructed by drawing a string of eight carbon atoms, numbering them 1 through 8, completing a ring by connecting the last atom to atom 3 (C8-3), backtracking to atom 1 and drawing a string of five carbon atoms attached to it, closing a second ring by connecting the last atom to atom 1 (1C5-1), numbering the added atoms 9 through 13, and adding the substituents. The substituent data (including hydrogen) precede the ring data and in each case the substituent data include the ring atom. Thus, a hydrogen atom is attached to each of carbons 4, 5, and 8; a chlorine atom is attached to carbon 11, and so on.

3.4 Development of Graph Theory

The development of graph theory is described in detail elsewhere in this book, so its development is traced only briefly here as it applies to chemical nomenclature.

The first recorded use of graph theory was the resolution of the Koenigsberg bridge problem by Euler [54] in 1736. Since then, graph theory has been applied to such diverse subjects as civil and electrical engineering, economics, distribution networks, and to a variety of chemical subjects [55].

Graph theory was first applied to a chemical problem by Cayley when he proposed the concept of the tree [56] in 1857 and subsequently applied the concept to enumeration of hydrocarbons [57] in 1874. Since then, graph theory has been applied to a range of chemical subjects such as reaction graphs, synthesis design, chemical documentation, and kinetics. However, relatively little work has been done until recently on application

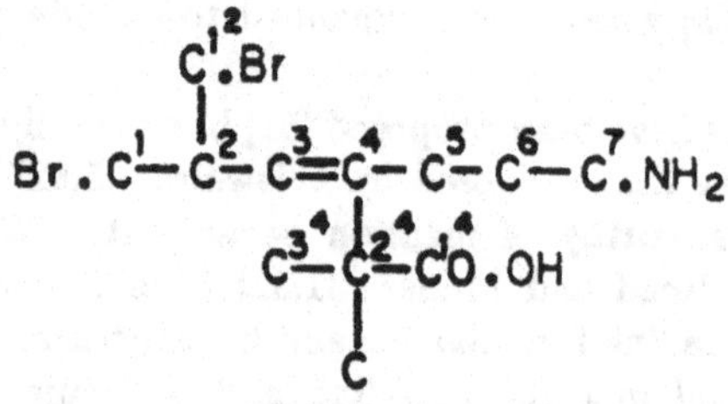

$$C^1.C^2Br$$

$$Br.C^1 - C^2 - C^3 \equiv C^4 - C^5 - C^6 - C^7.NH_2$$

$$C^3 - C^2 - C^4O.OH$$

$$C$$

C(CH:CH₂)(= NH).CH₂.C(Cl₃):CH.CH(CH₃)₂

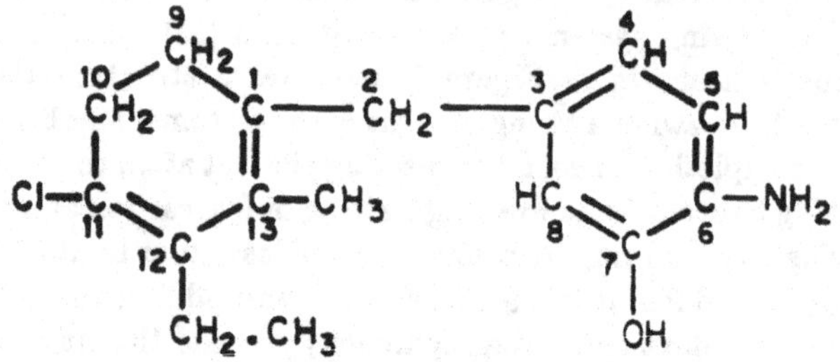

$$^{10}CH_2 \quad ^9CH_2 \quad ^1C - ^2CH_2 - ^3C \quad ^4CH \quad ^5CH$$

$$Cl - ^{11}C = ^{12}C \quad ^{13}C - CH_3 \quad HC - ^8 \quad ^7C - ^6C - NH_2$$

$$CH_2.CH_3 \quad OH$$

4,5,8-CH;11-CCl,2,9,10-CH₂;7-C(OH);13-C(CH₃);
6-C(NH₂);12-C(CH₂.CH₃);C8-3,1C5-1

Figure 5. Read's notation.

of graph theory to chemical nomenclature.

For the chemist unfamiliar with graph theory, a number of publications cover its application to chemistry [58–61], such as enumeration of isomers, topology, organometallic chemistry, and polymers. A list of definitions collected by Essam and Fisher [62] provides a useful introduction to the terminology of graph theory.

3.5 Application of Graph Theory to Chemical Nomenclature

Chemical nomenclature proposals based on hydrocarbons are of limited applicability because they cannot be used directly for chemical structures, such as boron cages and metallocenes, which contain atoms of valency (or connectivity) greater than 4 and for which there are no corresponding hydrocarbons. However, the concepts contained in such proposals can often be extended to graph-based nomenclature. The application of graph theory to chemical nomenclature has been reviewed by Goodson [39,40].

Conceptually, for both cyclic and acyclic structures, there are two approaches to numbering and naming chemical graphs. For ring systems, the approaches are (i) to pick a path through each ring system, and (ii) to number and name each component ring in turn. For acyclic systems, they are (i) to consider each system as one unit, and (ii) to consider each system as the sum of unbranched components. Numbering of atoms or nodes is of fundamental importance to any nomenclature but is not always taken into consideration. Unique numbering of atoms or nodes is preferable because it eliminates ambiguity in, for example, description of stereochemistry.

Of the nomenclature proposals bassed on hydrocarbons to which graph theory can be applied, the one developed by von Baeyer [63] encompasses the approach of picking a path through ring systems, as illustrated in Figure 6. In principle, two carbon atoms are joined by three bridges: *a*, *b*, and *c*. The numbers of atoms in the bridges are listed in ascending order and enclosed in brackets in the name.

As discussed by Goodson [64], subtraction of 1 from the number of terms inside the brackets of the first example of Figure 6, which is 3, yields the number of rings, which is 2. In other words, the numbers inside the brackets are a mathematical description of the graph of the structure. This can be taken further by summing the numbers inside the brackets (i.e., $0 + 1 + 4$), which gives 5, and adding 2 for the number of bridgeheads to yield 7, the number of nodes in the graph.

The ability to calculate the number of rings and nodes from the data inside the brackets means that some terms (e.g., "bicyclo" and "hept" in this example) could be regarded as redundant though, of course, they are needed for indexing purposes. The only information in a von Baeyer name not related to the data inside the brackets is the German ending "an" ("ane" in English) which, by convention, implies singly-bound carbon atoms. Von Baeyer's proposal, with minor modifications to the format, has been generally adopted and extended [17a] to cyclic hydrocarbons with more than two rings and/or with heteroatoms, as illustrated in Figure 6.

Balaban [65] has also discussed the mathematical aspects of the von Baeyer nomenclature. He went on to discuss its problems; i.e., that it is cumbersome, that there is redundancy in the names, and that deriving

Von Baeyer's proposal

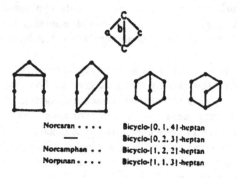

Norcaran	Bicyclo-[0. 1. 4]-heptan
—	Bicyclo-[0. 2. 3]-heptan
Norcamphan . .	Bicyclo-[1. 2. 2]-heptan
Norpinan	Bicyclo-[1. 1. 3]-heptan

Extended von Baeyer nomenclature

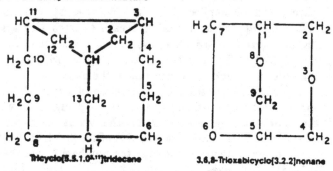

Tricyclo[5.5.1.0²·¹¹]tridecane

3,6,8-Trioxabicyclo[3.2.2]nonane

Figure 6. Von Baeyer ring nomenclature and its extension.

the correct names for the more complex ring systems is difficult and error-prone. Balaban then suggested replacing von Baeyer nomenclature by a notation. Alternatively, he suggested replacing von Baeyer nomenclature by one based on a convention for unambiguously cutting the graph of a ring system to convert it to a tree (as is done to determine the number of rings in a ring system [17b,19a,66]), then coding unambiguously from the center or centroid of the tree.

Taylor [29,67] adopted the alternative approach of numbering and naming component rings in his proposal, which is illustrated in Figure 7. Again, replacement nomenclature was used for heteroatoms. Taylor observed that "in the field of ring structures there is no written word or speech equivalent of the structural formula — names must be associated with the structural formula by sight". For example, there is nothing in the names "pyridazine", "pyrimidine", and "pyrazine" to indicate that they are the 1,2-, 1,3-, and 1,4-diaza analogs, respectively, of benzene. Taylor therefore attempted to develop "a nomenclature that is integrated with classification". His Latin–Greek combination names are written word or speech equivalents of structural formulas in that they express the number and size of the constituent rings and are, in fact, a verbal form of what is known either as the smallest set of smallest rings (SSSR) [68-70] or as the ring size part of ring analysis [71]. Thus, 7(12)-tria-1,6,8(13)-ternipenta-4,12(15)-binihexalane (Figure 7) provides a direct correlation between the SSSR of a ring system and its name. The ring system is saturated (lane) and contains one 3-membered (tria), three 5-membered (ternipenta), and two 6-membered rings (binihexa). When the example shown is drawn from the name, the locant 1 indicates that a 5-membered ring is drawn first and numbered. The next locant (i.e., 4) indicates that a 6-membered ring is fused to the first ring at bond 4-5 and numbered. Then, in turn, a 5-membered ring is fused at bond 6-7, a 3-membered ring at bonds 8-7 and 7-13, and finally a 6-membered ring at bonds 12-13 and 13-15.

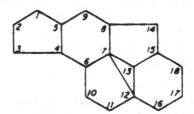

7(12)-Tria-1,6,8(13)-ternipenta-

4,12(15)-binihexalane

Figure 7. Taylor's ring nomenclature.

By directly correlating the SSSR of a ring system and its name, Taylor nomenclature lends itself directly to both manual and machine processing and raises the possibility of at least partial name assignment by computer.

While exploring the extension of Taylor's nomenclature to the graphs of complex ring systems, Goodson [72] proposed a systematic, mathematically-derived modification of Taylor's Latin and Greek number names which eliminated such irregularities as "undeni" for eleven versus "undeviceni" for nineteen versus "vicenisinguli" for twenty-one. The result of the modification can be seen in the examples illustrated in Figure 8. In these examples, Goodson has used Taylor's concepts to name graphs with nodes of connectivity greater than four. The rings are numbered as before, but "genisinguli" is used instead of "undeni" for eleven. Goodson also proposed the ending "gon" to differentiate the names of graphs from those of hydrocarbons.

1(5),1(6),1(8),2(4),3(7),2(7),
4(10),9(13),2(9),11,11(14)-
Genisinguiltetra-1-pentagon

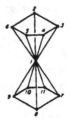

1,1(3),1(4),1(5),1(1),1(6),
1(9),1(10),2(6),7(11)-
Genitrigon

Figure 8. Extension of Taylor's nomenclature to ring graphs with nodes of connectivity greater than four.

A comprehensive organic nomenclature system was proposed by Dyson [35] in which acyclic structures are considered as the sum of unbranched components and component rings of ring systems are numbered in turn. The Taylor and Dyson nomenclatures contain similarities but were developed independently [49a]: a major difference is that numbering of ring systems begins at a ring junction in the Dyson system. Dyson found that it had proved "possible to form a basic 'mathematical' language for chemical structures, at least insofar as structures entirely constituted from carbon and hydrogen are concerned". This is reflected in the ring system names which, like Taylor names, are a verbal form of the SSSR, as illustrated in Figure 9. Thus, the name of the first example describes a ring system containing two 8- (biocta), two 6- (bihexa), and two 5-membered saturated rings (bipentalan). The locants at the end of the name define how the rings are fused and they occur in the same order as the ring sizes in the name with the exception of the first (8-membered) ring, which is assumed. Thus, locant 1 indicates that a 6-membered ring is fused to the assumed 8-membered ring at bond 1-2, then locants 3-12 indicate that a 5-membered ring is fused to bonds 3-2 and 2-12. The process is continued until the ring system is complete. The name of the second example (from which hydrogen is omitted) describes an acyclic structure with a main chain of seven carbon atoms (heptan) and a double bond between atoms 3 and 4 (en-3). To atom 4 is attached a 3-carbon branch through its atom 2 (2-trian) to which is attached a 1-carbon secondary branch (monan-2). One of the terminal atoms of the main branch is part of a carboxylic acid group (acid). A monobrominated single-carbon branch is attached to atom 2 of the main chain ((monan,brom)2), and NH_2 group is attached to atom 7 (azan-7) and a bromine atom to atom 1 of the main chain (brom). Thus, locant numbering in this nomenclature is not unique.

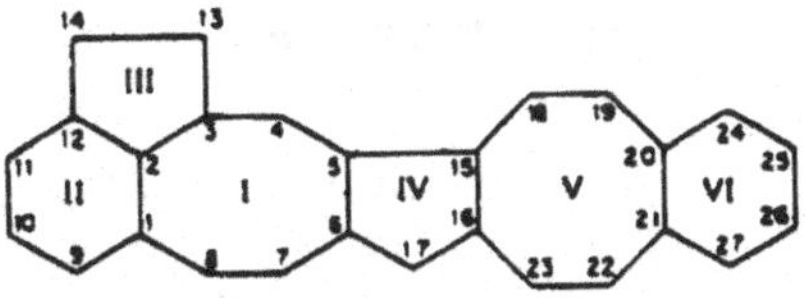

Biocta,bihexa,bipentalan,15,1,20,3-12,5

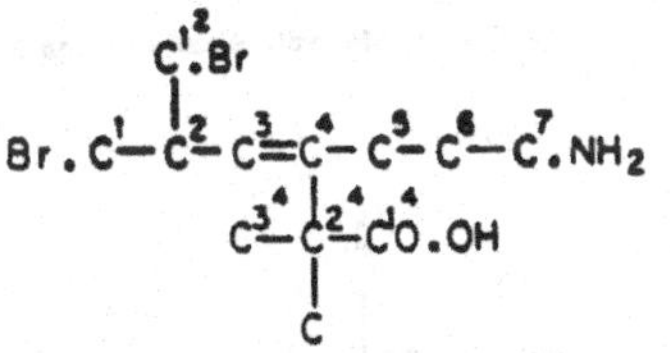

Heptan,en-3(2-trian,monan-2,acid)4,
(monan,brom)2,azan-7,brom

Figure 9. Dyson's nomenclature.

Another comprehensive organic nomenclature was proposed by Terent'ev *et al* [30]. Ring systems are numbered uniquely by picking a path, and acyclic compounds are numbered and named uniquely as sums of unbranched components. The name of the first example in Figure 10 is similar to a von Baeyer name. The numerical data are read as follows.

The longest path contains thirteen atoms (13). Atom 14 is not connected to atom 13 but bridges atoms 11 and 8 (x14(11,8)). Atom 13 is connected to atom 5 (13-5) and atom 12 is connected to atom 7 (12-7). The second example illustrates the use of replacement nomenclature (aza), of "arene" to represent the benzene ring, and the naming of all functional groups as suffixes, in contrast to IUPAC practice [17c]. The third example illustrates the use of letters and primes to provide unique numbering of acyclic structures, naming of branched groups as simple radicals, and grouping of like radicals in a name.

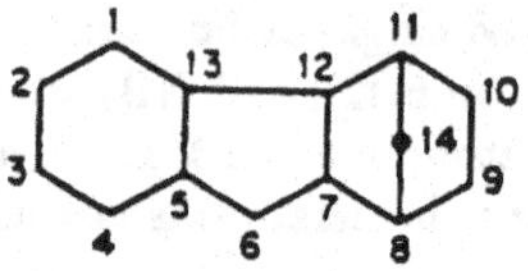

Tetracyclo[13.x14(11,8); 13-5; 12-7]tetradecane

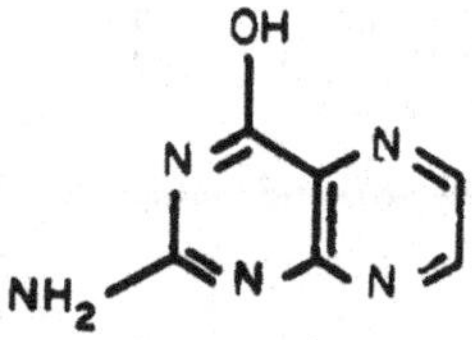

1,3,6,9-Tetraazadecadiarenol-4-amine-2

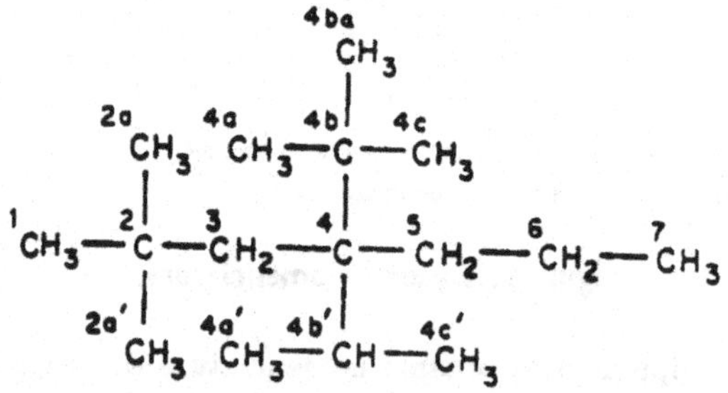

2,2,4b-Trimethyl-4,4-di-2-propylheptane

Figure 10. The nomenclature of Terent'ev *et al.*

Balaban [33,34,60] has proposed a nomenclature system for poly-cyclic aromatic hydrocarbons (arenes), which are superimposable on the graphite lattice. The nomenclature is based on their characteristic (or, better, dualist [73]) graphs, which are defined as the connected graphs obtained by joining the centers of the component benzene rings. If the connected graph of an arene is a tree, the arene is *cata*-condensed; if the graph contains rings, the arene is *peri*-condensed. As illustrated in Figure 11, Balaban's proposal greatly reduces the diversity of names for these arenes (0, 1, and 2 mean straight on, left, and right, respectively, and (•) indicates a branch). Balaban has extended this proposal, as a notation only [61a,74], to structures containing rings larger and smaller than the benzene ring. No method of numbering atoms or nodes has been provided, though, from the nature of the nomenclature, numbering each component ring in turn would appear to be suitable.

Elk [22] has proposed a nomenclature for arenes (arene being defined to include ring sizes other than six) which incorporates graph theoretical analysis techniques and is a refinement of his earlier proposal [75] based on geometry. In this system, pentalene, indene, and naphthalene are named 5,5-, 6,5-, and 6,6-diarene, respectively, and for larger ring systems letters in the names indicate where additional rings are fused. Thus, anthracene, phenanthrene, and phenalene become 6,6,6a-, 6,6,6b-, and 6,6,c-triarene, respectively. The process is repeated for additional rings as shown in Figure 12, so that naphthacene, chrysene, and pyrene are named 6,6,6a,6a-, 6,6,6b,6b-, and 6,6,6b,6e-tetraarene, respectively. Like Balaban, Elk does not include a scheme for numbering individual atoms or nodes but, again, numbering each component ring in turn appear to be suitable.

Balaban and Schleyer [73,76] have applied the dualist graph princi-ples developed for arenes to propose a nomenclature system for polyman-tanes; i.e., hydrocarbons that contain at least one adamantane unit and are superimposable on the diamond lattice. When the centers of fused adamantane units in polymantanes are joined, a dualist graph results that is also superimposable on the diamond lattice. As illustrated in Figure 13, the four tetrahedral directions of the dualist graph are designated by the numerals 1 through 4. These numerals are enclosed in brackets in the names of the polymantanes, indicating the relative bonding directions with branches being indicated by parentheses. The term tetramantane in each example indicates that the ring systems are composed of four adamantane units. The proposal yields much simpler names for the three tetramantanes, the von Baeyer names for which are quite cumbersome.

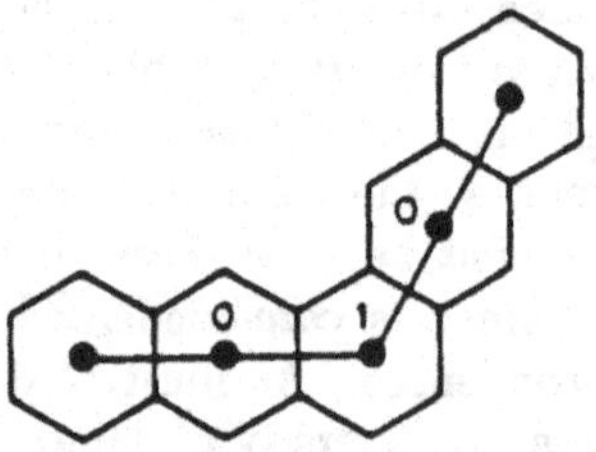

(010)Pentacatafusene
(Pentaphene)

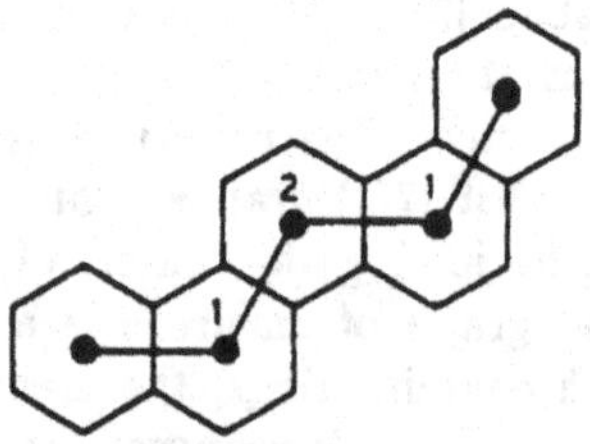

(121)Pentacatafusene
(Picene)

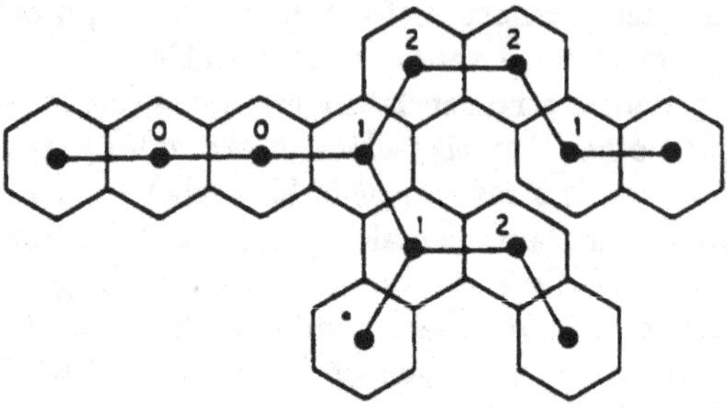

(001[1(.)2]221)Dodecacatafusene

Figure 11. Balaban's nomenclature for *cata*-condensed polycyclic aromatic hydrocarbons.

No procedure for numbering atoms or nodes was described, though, again, numbering each component ring in turn would appear to be suitable.

The only comprehensive graph-based nomenclature system proposed to date is that developed by Lozac'h *et al* [38,77] and called "nodal nomenclature" (for which an algorithm is being written), from the nodes representing atoms in graphs. Nodal nomenclature is not limited to organic substances but can also number and name coordination and other structures. It also uniquely numbers a ring system by picking a path, and a branched acyclic structure is considered as one unit. In general terms, a chemical structure is named by first identifying the atoms or groups to be treated as substituents, then deriving and naming the corresponding graph, and then naming the chemical structure as a derivative of the

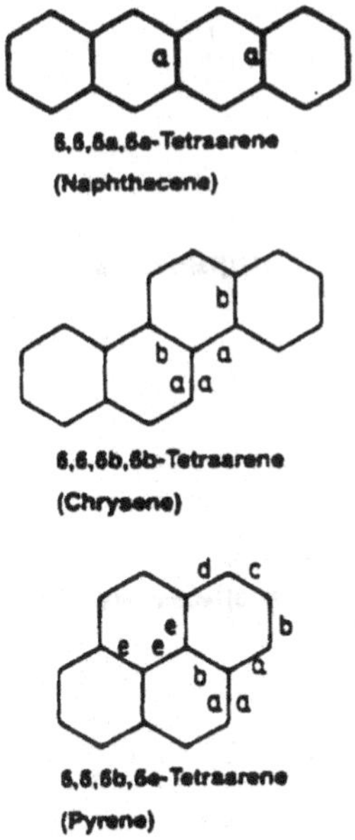

6,6,6a,6a-Tetraarene
(Naphthacene)

6,6,6b,6b-Tetraarene
(Chrysene)

6,6,6b,6a-Tetraarene
(Pyrene)

Figure 12. Elk's nomenclature for polycyclic aromatic hydrocarbons.

graph, to yield names such as those as illustrated in Figure 14. The name of the first example describes an unbranched chain of six carbon atoms ([6]hexan), two of which are replaced by nitrogen atoms (2,5-diaza). Carbon 6 is part of an acid group (6-(oxido,oxo)) and a proton has been added to nitrogen 2 (2-ium). In the second name, the descriptor (i.e., the numerals in brackets) is a complete mathematical description of the graph. Thus, a ring of ten atoms is bridged from atom 1 to atom 6 ((010.0^{1,6})) to yield two rings, which is confirmed by the term "bicyclo"; this module is connected through atom 3 to atom 11 of the second module (3:11), which is a chain of three atoms (3); and so on. Summing the number of atoms (10+3+2+1) yields sixteen which is confirmed by the term "hexadecan". Atoms 1 through 10 are connected by aromatic bonding ((1-10)arene) and atoms 13, 15, and 16 are each part of an acid group (13,15,16-trioic acid). Although the substitutive method can be used in nodal nomenclature, its use is significantly different from that recommended by IUPAC [17d] because of the nature of nodal nomenclature. This is the first time that graphs have been named as a means of naming chemical

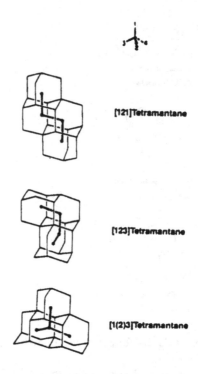

[121]Tetramantane

[123]Tetramantane

[1(2)3]Tetramantane

Figure 13. Polymantane nomenclature of Balaban and Schleyer.

structures (cf. Goodson [72], above).

The MOST-3SR nomenclature system of Iizuka *et al* [78,79] for polycyclic hydrocarbons is based on the Morgan labeling of spanning trees (MOST), the concept of the smallest set of smallest rings (SSSR or 3SR), and the concept of numbering component rings in sequence. Its algorithm derives names such as those illustrated in Figure 15 by assigning Morgan numbering, finding the smallest set of smallest rings, and renumbering the component rings in a prescribed order. The numerical data in each name show the ring sizes in the order they are numbered (e.g., 5,6,5,5,5), then, except for the first ring, the higher locant of the preceding ring or ring system to which the next ring is attached, the number (in parentheses) of atoms required to complete the new ring, and the lower locant to which the new ring is attached (e.g., 2(4)1).

While the graph center concept has been known for over 100 years

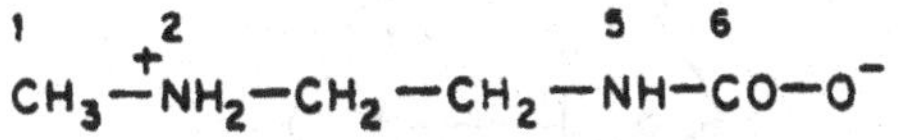

$$\overset{1}{CH_3}-\overset{\overset{2}{+}}{NH_2}-CH_2-CH_2-\overset{5}{NH}-\overset{6}{CO}-O^-$$

6-(Oxido,oxo)-2,5-diaza[6]hexan-2-ium

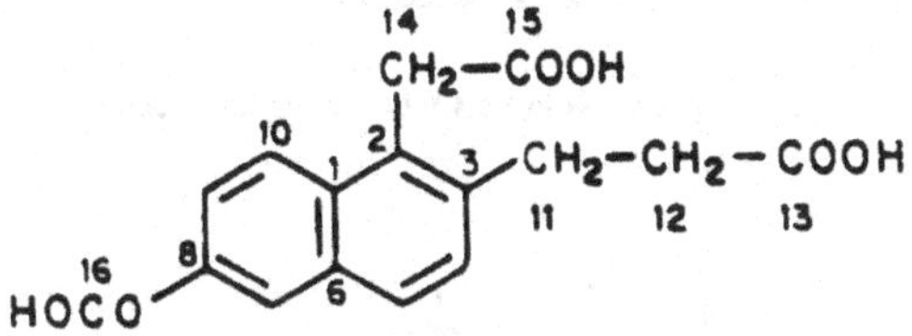

Bicyclo[(010.0¹⁻⁹)3:11(3)2:14(2)8:16(1)]hexadecan
(1-10)arene-13,15,16-trioic acid

(6-Carboxy-1-(carboxymethyl)-2-naphthalene propionic acid)

Figure 14. Nodal nomenclature.

in connection with acyclic graphs [80], it is only recently that it has been extended to polycyclic graphs and used as the basis for a nomenclature [81,82] for polycyclic hydrocarbons (cf. [65]). This proposal is not limited to *cata*-condensed benzenoid systems, as was an earlier proposal [33,34]. The benzenoid rings in the graphite (honeycomb) lattice are given standard numbering, as shown in Figure 16a, with the central (or focal) point being numbered 0. The successive shells around the focal point and the standard numbering of the vertexes of the dualist graphs are shown in Figure 16b. This numbering is used to indicate in the name the relative positions of the component rings. The slashes (/) in the names define in which shells the rings occur. Thus, the first two examples in Figure 13 start with a ring at the focal point and have two (1,4) and three (1,2,3) rings, respectively, in the first shell. The first example has two rings (2,8) in the second shell while the second example has one (1). The third example illustrates a ring system in which there is no ring at the focal point: this is indicated by a slash at the beginning of the name. No system of numbering the atoms or nodes was included with this proposal, though numbering each component ring in turn would appear to be suitable.

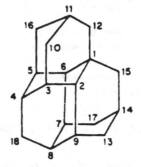

Pentacyclo-5,6,5,5,5:2(4)1,9(1)4,
7(3)6,12(1)6-tetradecane

(Pentacyclo[8.2.1.1⁴·⁷.0²·⁶.0⁶·¹²]tetradecane)

Heptacyclo-6,6,6,6,6,6,6:6(3)2,3(3)1,
9(3)1,11(1)5,14(1)7,8(1)4-octadecane

(Heptacyclo[7.7.1.1³·¹⁰.0¹·¹².0²·⁷.
0⁶·¹³.0⁶·¹¹]octadecane)

Figure 15. MOST-3SR nomenclature.

Bonchev [83] has recently extended the use of the graph center concept to propose a more comprehensive nomenclature system for organic compounds. Polycyclic benzenoid hydrocarbons (named "fuseenes"; the saturated analogs are "fusanes") are superimposed on the standard hexagonal lattice for numbering purposes, as before [82]. However, in this proposal not only are the rings numbered but also the atoms and bonds (Figure 17). Standard lattices are also used for acyclic structures; structures with odd-numbered chains have graphs with single centers while those with even-numbered chains have bicenters. Unlike numbering of

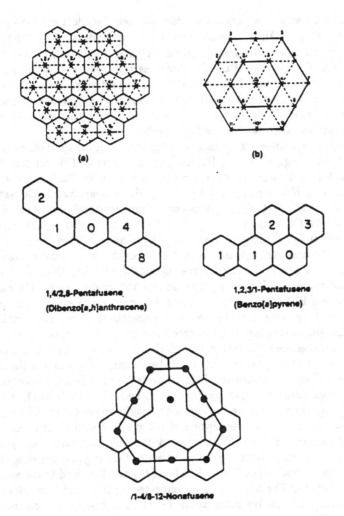

Figure 16. Nomenclature of polycyclic aromatic hydrocarbons based on topological centric coding.

cyclic structures, numbering of acyclic structures is not unique. However, differentiation is achieved by the use of slashes, as in "//1-ine" which indicates bond locant 1 in the second shell. Functional groups are listed at the end of a name in alphabetical order of their chemical symbols (e.g., 1-COOH, NH$_2$, 6-OH). Replacement atoms are indicated by their element symbols after the locants indicating their positions in the lattice (e.g., 1-Si). The nomenclature also indicates ring sizes, e.g., tetro, pento, and hexo for 4-, 5-, and 6-membered rings, respectively. Two additional features of this nomenclature are the naming of ring systems connected by acyclic bonds (bridganes) and the naming of ring clusters connected by ring fusions, spiro fusions, and by acyclic bonds.

Bonchev's nomenclature incorporates the chemical symbols of functional groups (e.g., COOH, NH$_2$, OH) in the name and it can therefore be regarded as a mixed nomenclature/notation. A similar proposal has been made by Krishnamurthy [26] for organic nomenclature. His system is named WISENOM, from "Wiswesser Nomenclature", and is derived from his ALWIN proposal [84,85]. WISENOM is based on graph coding and a context-free formal grammar (cf. [25]). The algorithm described encodes acyclic and cyclic structures and decodes the corresponding names. To encode an acyclic structure, the algorithm first identifies the main chain and branches. It then assigns locants and finally assigns the name from the WISENOM syntax, the syntax being in Backus–Naur form [86]. Cyclic structures are named by first identifying the smallest set of smallest rings, choosing a starting ring and successive rings for synthesis, assigning locants, synthesizing the system and deriving a unique code, and finally recording ring atoms, saturation, and substituents. The name of the first example in Figure 18 indicates a heteroatomic (*he*) acyclic structure (*tree*) of ten atoms (*dec*) with (at positions *b*, *e*, and *h*) three (*terni*) oxygen atoms (*oxa*). The structure contains only single bonds (*ane*). The punctuation = indicates that description of the main chain is complete. The name of the second example describes a carbonaceous (*ca*) ring system (*ring*). A quasi ring is one formed as a result of closing another ring. The second ring is fused to the first at bond *ab* (*/a*) and the third to the second at bond *hi* (*/h*). The fourth ring is connected to the third through spiro atom *j* and contains five other atoms (*j5j*). The first ring is completed by connecting atoms *g* and *p* by a single atom and by connecting atoms *h* and *q* directly (*glp,h-q*). Atoms *j* and *h* are two spiro atoms (*j, h*-bino spiro). The name of the third example can be similarly interpreted, the punctuation = indicating completion of the description of each fragment of the structure.

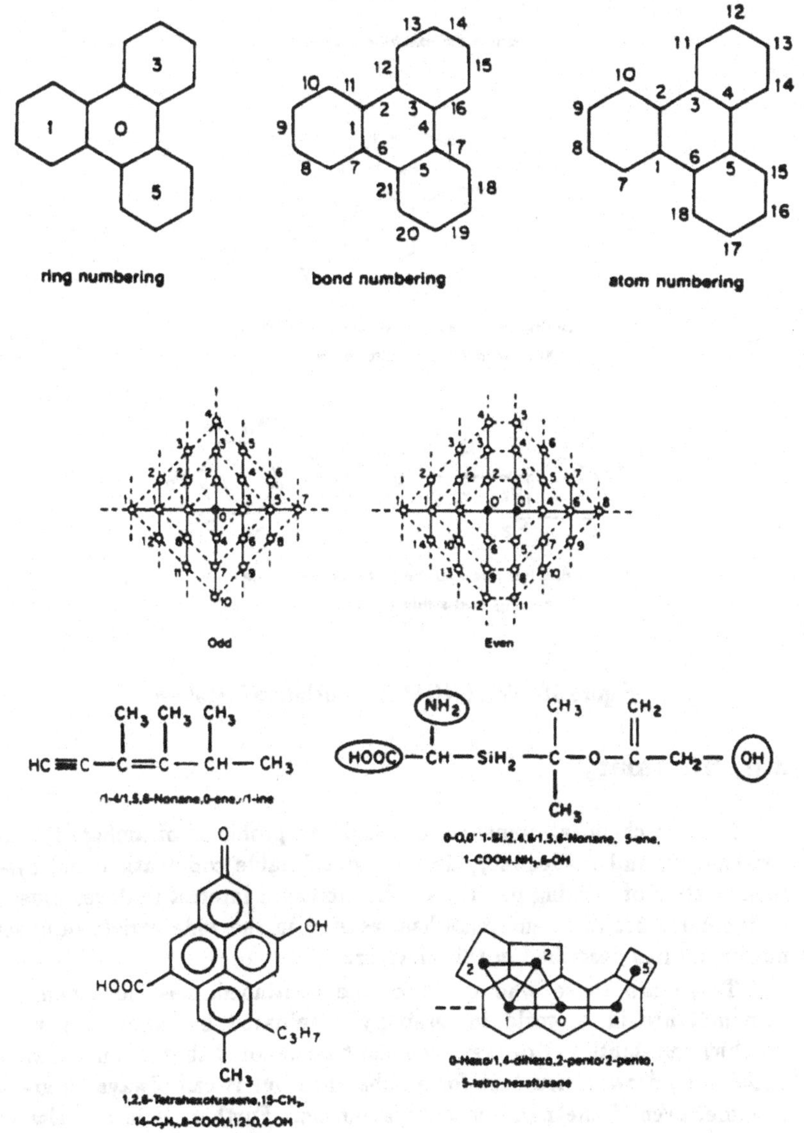

ring numbering   bond numbering   atom numbering

Odd   Even

Figure 17. Bonchev's organic nomenclature/notation based on the graph center concept.

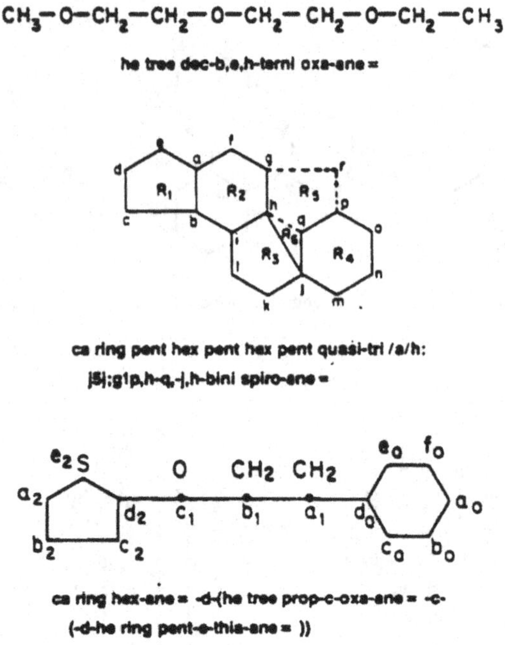

$$CH_3-O-CH_2-CH_2-O-CH_2-CH_2-O-CH_2-CH_3$$

he tree dec-b,e,h-terni oxa-ane =

ca ring pent hex pent hex pent quasi-tri /a/h:

|S|:g1p,h-q,-j,h-bini spiro-ane =

ca ring hex-ane = -d-(he tree prop-c-oxa-ane = -c-

(-d-he ring pent-e-thia-ane =))

Figure 18. WISENOM nomenclature/notation.

3.6 Summary

Modern chemical nomenclature still has problems of ambiguity, inconsistency, and inadequacy, despite considerable codification and systematization of existing practices. The increased interest in development of nomenclature in recent years has resulted in the wide variety of novel nomenclatures described in this chapter.

There are those who question the continued need for chemical nomenclature in a world increasingly dominated by computers with graphics capabilities. However, we must bear in mind that it is not always possible to draw a structure for a substance but it can always be given a name, even if the name is nonsystematic. Further, it is not always convenient to communicate structural information. So it appears likely that there will always be a need for nomenclature.

It is becoming increasingly apparent that in the future nomenclature will need to be capable of use by both the chemist and the computer. If such a nomenclature can be more comprehensive than existing practices, its usefulness will be increased because of the reduced need to choose from among a variety of practices (some of them overlapping) and the reduced number of rules to be learned or incorporated in a computer algorithm.

It is also becoming increasingly apparent that graph theory is providing a stimulus to development of novel nomenclatures that possess the required characteristics. This area will be a fruitful field for research for some time to come.

3.7 References and Notes

1. M.L. Huber, "Chemical Structure Codes in Perspective," J. Chem. Doc. 5 (1965), 4–8.
2. J.E. Rush, "Status of Notation and Topological Systems and Potential Future Trends," J. Chem. Inf. Comput. Sci. 16 (1976), 202–10.
3. (a) *Webster's Third New International Dictionary*, G. & C. Merriam Co., Springfield, Mass. 1971; p. 1534. (b) *ibid.*, p. 1543.
4. G.M. Dyson, M.F. Lynch and H.L. Morgan, "A Modified IUPAC-Dyson Notation System for Chemical Structures," Inform. Storage Retr. 4 (1968), 27–83.
5. E.G. Smith and P.A. Baker, *The Wiswesser Line-Formula Chemical Notation (WLN)*, 3rd ed.; Chemical Information Management, Cherry Hill, New Jersey 1976; p. 281.
6. H.L. Morgan, "The Generation of a Unique Machine Description for Chemical Structures – A Technique Developed at Chemical Abstracts Service," J. Chem. Doc. 5 (1965), 107–13.
7. P.G. Dittmar, R.E. Stobaugh and C.E. Watson, "The Chemical Abstracts Service Chemical Registry System. I. General Design," J. Chem. Inf. Comput. Sci. 16 (1976), 111–21.
8. R. Lees and A.F. Smith (Ed.), *Chemical Nomenclature Usage*, Ellis Horwood, Chichester 1983.
9. N.A. Farmer and M.P. O'Hara, "CAS ONLINE. A New Source of Substance Information from Chemical Abstracts Service," Database (12) (1980), 10–25.
10. C.R. Zeidner, J.O. Amoss and R.C. Haines, "The CAS ONLINE Computer Architecture for Substructure Searching," *Proc. 3rd Nat. Online Mtg.* 1980, 575–86.

11. M.P. Crosland, *Historical Studies in the Language of Chemistry*, Dover, New York 1978.
12. D.H. Rouvray, "The Chemical Applications of Graph Theory," MATCH (1) (1975), 61–70.
13. D.H. Rouvray, "The Changing Role of the Symbol in the Evolution of Chemical Notation," Endeavour 1 (1977), 23–31.
14. R. Hoppe, "On the Symbolic Language of the Chemist," Angew. Chem. 92 (1980), 106–21; Angew. Chem. Int. Ed. Eng. 19 (1980), 110–25.
15. For a comprehensive history of the development and activities of this organization, see [45], references to Verkade in [39], and P.E. Verkade, "Études Historiques sur la Nomenclature de la Chimie Organique. XVI. Historique de la Section D de l'I.U.P.A.C. Nomenclature of Organic Chemistry," Bull. Soc. Chim. Fr. (5–6, Pt. 2) (1979), 215–29.
16. IUPAC, *Nomenclature of Inorganic Chemistry*, 2nd ed., Definitive Rules 1970; Butterworths, London 1971.
17. IUPAC, *Nomenclature of Organic Chemistry: Sections A, B, C, D, E, F and H*, Pergamon, Oxford 1979.   (a) *ibid.*, Rules A-31, A-32, and B-14.   (b) *ibid.*, Rule A-32.12.   (c) *ibid.*, Rule C-10.   (d) *ibid.*, Rules C-10.1 through C-16.36.
18. An early example of the involvement of CAS in the development of nomenclature is the recognition in 1921 of the need for a catalog of ring systems, which eventually resulted in the publication of A.M. Patterson and L.T. Capell, *The Ring Index*, American Chemical Society, Washington 1940; 2nd ed., 1960.
19. To ensure that new substances are named consistently as they are reported, CAS has for many years had its own set of nomenclature rules which follow IUPAC practice as closely as possible. The most recent version of the rules is *Chemical Substance Name Selection Manual*, Chemical Abstracts Service, Columbus 1982.   (a) *ibid.*, Rule B-051.
20. A condensed version of the nomenclature rules published in [19] is published as "Chemical Substance Index Names," Appendix IV, *Chemical Abstracts 1982 Index Guide*, Chemical Abstracts Service, Columbus. It can be obtained separately as *Naming and Indexing of Chemical Substances for CHEMICAL ABSTRACTS*, Chemical Abstracts Service, Columbus 1982.
21. The references appearing in [20] as Section J can be obtained separately as *Selective Bibliography of Nomenclature of Chemical*

Substances, Chemical Abstracts Service, Columbus 1982.

22. S.B. Elk, "Refinement of Systematic Nomenclature for Polybenzenes and its Extension to Systems of General Arenes," MATCH (**13**) (1982), 239–69.

23. J. Mockus, A.C. Isenberg and G.G. Vander Stouw, "Algorithmic Generation of Chemical Abstracts Index Names. 1. General Design," J. Chem. Inf. Comput. Sci. 21 (1981), 183–95.

24. G.G. Vander Stouw, C. Gustafson, J.D. Rule and C.E. Watson, "The Chemical Abstracts Service Chemical Registry System. IV. Use of the Registry System to Support the Preparation of Index Nomenclature," J. Chem. Inf. Comput. Sci. 16 (1976), 213–8.

25. O. Mestyanek and P. Réti, "An Attempt to Formalize the Chemical Nomenclature," MATCH (2) (1976), 45–50.

26. E.V. Krishnamurthy, "WISENOM. A Formal Organic Chemical Nomenclature System," J. Chem. Inf. Comput. Sci. 22 (1982), 152–60.

27. For a comprehensive description of the DARC system, see Dubois *et al* beginning with J.E. Dubois, D. Laurent and H. Viellard, "Le Système de Documentation et d'Automatisation des Recherches de Correlations. Principes Généraux," C.R. Acad. Sci., Paris, Ser. C 263 (1966), 764–7 and ending with J.E. Dubois and A. Panaye, "Système DARC. XXI. Théorie de Génération-Description. Représentation des Boranes et de leurs Dérivés," Bull. Soc. Chim. Fr. (**7-8**, Pt. 2) (1976), 1229–39. See also J.E. Dubois, "Principles of the DARC Topological System," Entropie 25 (1969), 5–13.

28. J.D. Scott, "The Need for Reform in Inorganic Chemical Nomenclature," Chem. Rev. 32 (1943), 73–97.

29. F.L. Taylor, "Enumerative Nomenclature for Organic Ring Systems," Ind. Eng. Chem. 40 (1948), 734–8.

30. A.P. Terentév, V.M. Potapov, A.N. Kost and A.M. Tsukerman, "On the Systematic Nomenclature of Organic Compounds," Vestn. Mosk. Univ. (**10**), No. 6, Ser. Fiz.-Mat. Estestv. Nauk No. 4 (1955), 97–134.

31. R.C. Read and R.S. Milner, "A New System for the Designation of Chemical Compounds for the Purpose of Data Retrieval. I. Acyclic Compounds," Report to the University of the West Indies (1968). Revised (1978) as Research Report CORR-78-42 of the Department of Combinatorics and Optimization, University of Waterloo, Ontario, Canada N2L 3G1.

32. R.C. Read, "A New System for the Designation of Chemical Compounds for the Purpose of Data Retrieval. I. Theoretical Preliminar-

ies and the Coding of Acyclic Compounds. J. Chem. Inf. Comput. Sci. In press.

33. A.T. Balaban and F. Harary, "Chemical Graphs. V. Enumeration and Proposed Nomenclature of Benzenoid Cata-Condensed Polycyclic Aromatic Hydrocarbons," Tetrahedron 24 (1968), 2505–16.

34. A.T. Balaban, "Chemical Graphs. VII. Proposed Nomenclature of Branched Cata-Condensed Benzenoid Polycyclic Hydrocarbons," *Tetrahedron* 25 (1969), 2949–56.

35. G.M. Dyson, "Some New Concepts in Organic Chemical Nomenclature," October 1976. Available from Dr. W.A. Warr of the Chemical Structure Association, c/o ICI Pharmaceuticals Division, Alderley Park, Macclesfield, Cheshire SK10 4TG, England.

36. J.E. Rush, "Handling Chemical Structure Information," Annu. Rev. Inf. Sci. Tech. 13 (1978), 209–62.

37. R.S. Cahn and O.C. Dermer, *Introduction to Chemical Nomenclature*, 5th ed., Butterworths, Woburn, Mass. 1979.

38. N. Lozac'h, A.L. Goodson and W.H. Powell, "Nodal Nomenclature. I. General Principles," Angew. Chem. 91 (1979), 951–64; Angew. Chem. Int. Ed. Engl. 18 (1979), 887–99.

39. A.L. Goodson, "Graph-Based Chemical Nomenclature. 1. Historical Background and Discussion," J. Chem. Inf. Comput. Sci. 20 (1980), 167–72.

40. A.L. Goodson, "Use of Graphs in Chemical Nomenclature," Congressus Numerantium 33 (1981), 15–38.

41. D. Bonchev, "Principles of a Novel Nomenclature of Organic Compounds," Pure Appl. Chem. 55 (1983), 221–8.

42. K.L. Loening, in *Kirk-Othmer Encyclopedia of Chemical Technology*, 3rd ed.; Wiley, New York 1981; Vol. 16, pp. 28–46.

43. J.H. Fletcher, "The Nomenclature of Organic Chemistry," J. Chem. Doc. 7 (1967), 64–7.

44. W.C. Fernelius, K. Loening and R.M. Adams, "Notes on Nomenclature," J. Chem. Educ. 48 (1971), 433–4.

45. P.E. Verkade, "Études Historiques sur la Nomenclature de la Chimie Organique. XV. L'Oeuvre de Siboni sur la Nomenclature de la Chimie," Bull. Soc. Chim. Fr. (1-2, Pt. 2) (1978), 13–31. This reference contains a complete bibliography of the work of G. Siboni and M. Perino.

46. G.M. Dyson, "A New Notation for Organic Chemistry," Royal Institute of Chemistry Lecture, published jointly with the Chemical Society and the Society of Chemical Industry, October 1946; p. 26.

47. G.M. Dyson, *A New Notation and Enumeration System for Organic Compounds*, Longmans, London 1947; 2nd ed. 1949.
48. G.M. Dyson, Research (London) 2 (1949), 104–15.
49. IUPAC, *A Proposed International Chemical Notation*, Longmans, London 1958. (a) *ibid.*, p. 1.
50. IUPAC, *Rules for IUPAC Notation of Organic Compounds*, Longmans, London 1961.
51. J.E. Ash and E. Hyde, *Chemical Information Systems*, Ellis Harwood, Chichester 1975.
52. W.J. Wiswesser, *A Line-Formula Chemical Notation*, Thomas Y. Crowell Co., New York 1954.
53. R.C. Read, "A New System for the Designation of Chemical Compounds for the Purpose of Data Retrieval. II. Cyclic Compounds," Research Report CORR 80-7 of the Department of Combinatorics and Optimization, University of Waterloo, Waterloo, Ontario, Canada N2L 3G1.
54. J.R. Newman, "Leonhard Euler and the Koenigsberg Bridges," Sci. Am. 189 (1953), 66–70; translation of Comment. Acad. Sci. I Petropolitanae 8 (1736), 128.
55. R.J. Wilson and L.W. Beineke (Eds.) *Applications of Graph Theory*, Academic Press, London 1979. (a) D.H. Rouvray and A.T. Balaban, *Chemical Applications of Graph Theory*, ibid., pp. 177–221.
56. A. Cayley, "On the Theory of Analytical Forms Called Trees," Philos. Mag. 13 (1857), 172–6.
57. A. Cayley, "On the Mathematical Theory of Isomers," Philos. Mag. 67 (1874), 444–7.
58. D.H. Rouvray, "Graph Theory in Chemistry," RIC Rev. 4 (1971), 173–95.
59. D.H. Rouvray, "Uses of Graph Theory," Chem. Brit. 10 (1974), 11–8.
60. A.T. Balaban, "Some Chemical Applications of Graph Theory," MATCH 1 (1975), 33–60.
61. A.T. Balaban (Ed.) *Chemical Applications of Graph Theory*, Academic Press, London 1976. (a) *ibid.*, pp. 73–9.
62. J.W. Essam and M.E. Fisher, "Some Basic Definitions of Graph Theory," Rev. Mod. Phys. 42 (1970), 272–88.
63. A. Baeyer, "Systematik und Nomenclatur Bicyclischer Kohlenwasserstoffe," Ber. Dtsch. Chem. Ges. 33 (1900), 3771–5.
64. A.L. Goodson, "Application of Graph-Based Chemical Nomenclature to Theoretical and Preparative Chemistry," Croat. Chem. Acta 56

(1983), 319–28.

65. A.T. Balaban, "Chemical Graphs. XVIII. Graphs of Degree Four or Less, Isomers of Annulenes, and Nomenclature of Bridged Polycyclic Structures," Rev. Roum. Chim. 18 (1973), 635–53.

66. M. Frèrejacque, "Condensation d'une Molécule Organique," Bull. Soc. Chim. Fr. 6 (1939), 1008–11.

67. G.M. Dyson, "La Nomenclature des Hydrocarbures Polycycliques Système Dyson-Taylor-Patterson," Bull. Soc. Chim. Fr. (1957), 45–52.

68. M. Plotkin, "Mathematical Basis of Ring-Finding Algorithms in CIDS (Chemical Information and Data System)," J. Chem. Doc. 11 (1971), 60–3.

69. A. Zamora, "An Algorithm for Finding the Smallest Set of Smallest Rings," J. Chem. Inf. Comput. Sci. 16 (1976), 40–3.

70. B. Schmidt and J. Fleischhauer, "A Fortran IV Program for Finding the Smallest Set of Smallest Rings of a Graph," J. Chem. Inf. Comput. Sci. 18 (1978), 204–6.

71. "Ring Analysis Index, Index of Parent Compounds I," *Parent Compound Handbook*, Chemical Abstracts Service, Columbus 1976.

72. A.L. Goodson, "Graph-Based Chemical Nomenclature. 2. Incorporation of Graph-Theoretical Principles into Taylor's Nomenclature Proposal," J. Chem. Inf. Comput. Sci. 20 (1980), 172–6.

73. A.T. Balaban, "Enumeration of Catafusenes, Diamondoid Hydrocarbons, and Staggered Alkane C-Rotamers," MATCH (2) (1976), 51–61.

74. A.T. Balaban, "Chemical Graphs. XI. (Aromaticity. IX) Isomerism and Topology of Non-Branched *Cata*-Condensed Polycyclic Conjugated Non-Benzenoid Hydrocarbons," Rev. Roum. Chim. 15 (1970), 1251–62.

75. S.B. Elk, "A Nomenclature for Regular Tessellations and its Application to Polycyclic Aromatic Hydrocarbons," MATCH (8) (1980), 121–58.

76. A.T. Balaban and P.v.R. Schleyer, "Systematic Classification and Nomenclature of Diamond Hydrocarbons. I. Graph-Theoretical Enumeration of Polymantanes," Tetrahedron 34 (1978), 3599–609.

77. N. Lozac'h and A.L. Goodson, "Nodal Nomenclature. II. Specific Nomenclature for Parent Hydrides, Free Radicals, Ions and Substituents," Angew. Chem. In press.

78. T. Iizuka, M. Imai, N. Tanaka, T. Kan and E. Osawa, "Graph Theoretical Studies of Molecular Structures. IX. Searches for the

Neighbourhood of Pentagonal Dodecahedrane," Gunma Daigaku Kyoikugakubu Kiyo, Shizen Kagaku Hen 30 (1981), 5–12.

79. T. Iizuka, N. Tanaka, H. Miura and T. Kan, "A New Nomenclature for Polycyclic Hydrocarbons Based on the Smallest Set of Smallest Rings from the Morgan Labeling of the Spanning Trees," J. Chem. Inf. Comput. Sci. In press.

80. D. Koenig, *Theorie der Endlichen und Unendlichen Graphen*, Chelsea, New York 1950, p. 64.

81. D. Bonchev, A.T. Balaban and M. Randić, "The Graph Center Concept for Polycyclic Graphs," Int. J. Quant. Chem. 19 (1981), 61–82.

82. D. Bonchev and A.T. Balaban, "Topological Centric Coding and Nomenclature of Polycyclic Hydrocarbons. 1. Condensed Benzenoid Systems (Polyhexes, Fusenes)," J. Chem. Inf. Comput. Sci. 21 (1981), 223–9.

83. D. Bonchev, "Principles of a Novel Nomenclature of Organic Compounds," Pure Appl. Chem. 55 (1983), 221–8.

84. E.V. Krishnamurthy, P.V. Sankar and S. Krishnan, "ALWIN-Algorithmic Wiswesser Notation System for Organic Compounds," J. Chem. Doc. 14 (1974), 130–41.

85. S. Krishnan and E.V. Krishnamurthy, "Compact Grammar for Algorithmic Wiswesser Notation Using Morgan Name," Inf. Process. Manage. 12 (1976), 19–34.

86. J.W. Backus, "The Syntax and Semantics of the Proposed International Algebraic Language of the Zurich ACM-GAMM Conference," *Information Processing, Proceedings of ICIP Paris*, UNESCO, Paris 1960, pp. 125–32.

83. A.T. Balaban and P.v.R. Schleyer, "Systematic Classification and Nomenclature of Diamond Hydrocarbons. I. Graph-Theoretical Enumeration of Polymantanes," Tetrahedron 34 (1978), 3599–609.

84. N. Lozac'h and A.L. Goodson, "Nodal Nomenclature. .II. Specific Nomenclature for Parent Hydrides, Free Radicals, Ions and Substituents," Angew. Chem. 96 (1984), 1–15; Angew. Chem. Int. Ed. Engl. 23 (1984), 33–46.

85. R.C. Read and G. Hu, "An Algorithm for Labelling and Representing a Tree According to the Rules of Nodal Nomenclature," Research Report CORR 85-13 of the Department of Combinatorics and Optimization, University of Waterloo, Waterloo, Ontario, Canada N2L 3G1.

86. T. Iizuka, M. Imai, N. Tanaka, T. Kan and E. Osawa, "Graph Theoretical Studies of Molecular Structures. IX. Searches for the Neighbourhood of Pentagonal Dodecahedrane," Gunma Daigaku Kyoikugakubu Kiyo, Shizen Kagaku Hen 30 (1981), 5-12.
87. T. Iizuka, N. Tanaka, M. Imai, T. Kan and H. Miura, "Graph Theoretical Studies of Molecular Structures. XI. Application of the MOST-3SR Names to Spiro Alicyclic Hydrocarbons," Gunma Daigaku Kyoikugakubu Kiyo, Shizen Kagaku Hen 32 (1983), 17-25.
88. T. Iizuka, N. Tanaka, H. Miura and T. Kan, "A New Nomenclature for Polycyclic Hydrocarbons Based on the Smallest Set of Smallest Rings from the Morgan Labeling of the Spanning Trees," J. Chem. Inf. Comput. Sci. In press.
89. D. Koenig, *Theorie der Endlichen und Unendlichen Graphen*, Chelsea, New York 1950, p. 64.
90. D. Bonchev, A.T. Balaban, M. Randić, "The Graph Center Concept for Polycyclic Graphs," Int. J. Quant. Chem. 19 (1981), 61-82.
91. D. Bonchev and A.T. Balaban, "Topological Centric Coding and Nomenclature of Polycyclic Hydrocarbons. 1. Condensed Benzenoid Systems (Polyhexes, Fusenes)," J. Chem. Inf. Comput. Sci. 21 (1981), 223-9.
92. D. Bonchev, "Principles of a Novel Nomenclature of Organic Compounds," Pure Appl. Chem. 55 (1983), 221-8.
93. E.V. Krishnamurthy, P.V. Sankar and S. Krishnan, "ALWIN-Algorithmic Wiswesser Notation System for Organic Compounds," J. Chem. Doc. 14 (1974), 130-41.
94. S. Krishnan and E.V. Krishnamurthy, "Compact Grammar for Algorithmic Wiswesser Notation Using Morgan Name," Inf. Process. Manage. 12 (1976), 19-34.
95. J.W. Backus, "The Syntax and Semantics of the Proposed International Algebraic Language of the Zurich ACM-GAMM Conference," *Information Processing, Proceedings of ICIP Paris*, UNESCO, Paris 1960, pp. 125-32.
96. N. Lozac'h "Principles for the Continuing Development of Organic Nomenclature," J. Chem. Inf. Comput. Sci. 25 (1985), 180-5.

Chapter 4

POLYNOMIALS IN GRAPH THEORY

Ivan Gutman

Faculty of Science, University of Kragujevac, Yugoslavia

4.1 Why Polynomials in Graph Theory?

A variety of problems in pure and applied graph theory can be treated and solved in a rather efficient manner by making use of polynomials. Polynomials provide both a convenient and powerful mathematical tool and a valuable proof technique in certain fields of graph theory. There are essentially three routes by which polynomials enter into the theory of graphs.

First: Polynomials appear as generating functions of combinatorial graph invariants.

For example, let $m(G,k)$ be the number of ways in which k independent edges can be selected in the graph G. Then the sequence $m(G,1)$, $m(G,2)$, $m(G,3),\ldots$ etc., can be presented by means of the power series:

$$M(G,x) = 1 + m(G,1)x + m(G,2)x^2 + m(G,3)x^3 + \ldots,$$

with x being an auxiliary and meaningless variable. The above series is described as the generating function for the numbers $m(G, k)$. It will be seen later that the function $M(G, x)$ is in fact a polynomial.

A generating function can be viewed merely as shorthand notation for the respective sequence. There are, however, many advantages to using a generating function instead of a sequence of graph invariants. For instance, one can easily demonstrate that $M(G_{11}, x) = 1 + 3x$. The same generating function for the graph G_{12} can be calculated from the formula:

$$M(G_{12}, x) = (1 + 4x)^3\, M\Big(G_{11}, \frac{x}{(1 + 4x)^2}\Big).$$

$$G_{11} \qquad\qquad\qquad\qquad G_{12}$$

On the other hand, the direct determination of the sequence $m(G_{12},\ 1),\ n(G_{12}, 2), \ldots$ would be a considerably more laborious task.

Second: Polynomials can be used to introduce algebraic concepts into graph theory.

For example, the eigenvalues of the adjacency matrix are interpreted as the eigenvalues of the corresponding graph. As it is well known in linear algebra, in order to determine the eigenvalues one has to find the zeros of the characteristic polynomial. This has motivated a number of workers to introduce and extensively study the characteristic polynomial of a graph.

The importance of combining graph theory with algebra (via polynomials) is best illustrated by the fact that a recent book on the subject contains a bibliography comprising more than 650 items; since its publication some 400 additional contributions have been made to the field 154.

Third: The study of graphic polynomials is an interesting and relevant task *per se*.

For example, it is an intriguing and not at all obvious result that the generating function $M(G, x)$ defined above, although having a purely combinatorial origin, had a remarkable algebraic property: all its zeros are real numbers. Another curious finding is that the product of the zeros of the characteristic polynomial of a benzenoid graph is equal to the square of the number of perfect matchings of this graph.

4.2 On Chemical Applications of Graphic Polynomials

Graphic polynomials are frequently used in chemical applications of graph theory. The reasons for this should become clear after reading the present chapter. Here we shall briefly summarize the main directions in theoretical chemistry in which graphic polynomials are of some relevance.

Hückel molecular orbital theory and related simple one-electron models are concerned with the topology of conjugated molecules. The characteristic polynomial of the molecular graph functions in this context as the secular determinant. For more details on this topic see the chapter "Graph Theory and Molecular Orbitals" by Professor N. Trinajstić in this volume.

The theory of aromaticity, based on the so-called topological resonance energy, requires knowledge of the zeros of the matching polynomial.

Clar's theory of the aromatic sextet can be formulated in terms of the coefficients of the sextet polynomial.

The monomer-dimer model of statistical physics describes a simple system in which phase transitions can occur. It too makes use of the matching polynomial.

Hosoya's topological index predicts various physicochemical properties of saturated hydrocarbons from the coefficients of the Z-counting polynomial.

The coefficients of the characteristic, matching and sextet polynomials are related to the number of Kekulé and Dewar structures of the pertinent conjugated molecule. Hence all these polynomials are of some relevance in **resonance theory.**

The coefficients and zeros of the characteristic and matching polynomials reflect the extent of branching in the skeleton of a molecule. Accordingly, they have been used in designing several measures of branching and related **topological indices**. Several attempts have also been made to use graphic polynomials in chemical documentation. For more details on these topics see the chapters in later volumes of this series written by Bonchev, Mekenyan, Trinajstić and Rouvray.

Throughout the present chapter we shall use the notation and terminology as introduced by Professor O.E. Polansky in the chapter "Elements of Graph Theory for Chemists." We shall also deal exclusively with simple graphs and their graphic polynomials, and also discuss the relevant chemical applications. It should not be forgotten, however, that graphs with weighted edges and self-loops are of great importance in theoretical chemistry, for they represent heteroconjugated molecules. A great part of the theory which follows has been generalized to such graphs. For further details on the characteristic polynomials [2–8,181], matching polynomials [3,9,10] and quantum-chemical applications [11–14] of graphs with weighted edges and self-loops, the reader is invited to consult the literature cited.

4.3 Polynomials

Let $a_0, a_1, \ldots, a_{n-1},\ a_n$ be a set of numbers and let $a_0 \neq 0$. Then the expression:

$$P(x) = a_0 x^n + a_1 x^{n-1} + \cdots + a_{n-1} x + a_n,$$

which may also be written in the more compact form:

$$P(x) = \sum_{i=1}^{n} a_i x^{n-i},$$

is referred to as a polynomial in the variable x. The numbers $a_0, a_1, \ldots, a_n$ are the coefficients of the polynomial $P(x)$.

The degree of $P(x)$ is n (provided, of course, $a_0 \neq 0$).

If x_0 is a number which renders $P(x_0) = 0$, we describe x_0 as a zero of the polynomial $P(x) = 0$.

If $P(x)$ is a polynomial of degree n, $P(x)$ will have n zeros (which need not all be distinct). If these zeros are denoted by $x_1, x_2, \ldots, x_n$, then

$$P(x) = a_0(x - x_1)(x - x_2)\ldots(x - x_n).$$

The zeros of a polynomial are not necessarily real numbers.

For example, the simple polynomial $x^2 + 1$ has two zeros, but neither is a real number. Exercise: find all the zeros of the polynomials $P_1(x) = x^2 - 6$ and $P_2(x) = x^2 + 1$.

If $Q(x)$ and $R(x)$ are two polynomials and $P(x) = Q(x) \cdot R(x)$, the polynomial $Q(x)$ is said to be a divisor of the polynomial $P(x)$.

4.4 The Characteristic Polynomial

4.4.1 Theory

We recall first a few elementary notions from linear algebra. Let M be a square matrix of order n:

$$M = \begin{bmatrix} M_{11} & M_{12} & \ldots & M_{1n} \\ M_{21} & M_{22} & \ldots & M_{2n} \\ \cdot & \cdot & \ldots & \cdot \\ M_{n1} & M_{n2} & \ldots & M_{nn} \end{bmatrix}.$$

The characteristic polynomial of M is obtained by expanding the determinant of the matrix $(xI - M)$, with I being the unit matrix of order n and x a scalar variable.

The unit matrix has the property $MI = IM = M$ and assumes the form:

$$I = \begin{bmatrix} 1 & 0 & \ldots & 0 \\ 0 & 1 & \ldots & 0 \\ \cdot & \cdot & \ldots & \cdot \\ 0 & 0 & \ldots & 1 \end{bmatrix},$$

It is easy to see that the characteristic polynomial of M is a polynomial in the variable x of degree n.

The concept of the characteristic polynomial of a matrix plays an important role in linear algebra; the relevant theory is now well established. In particular, it is known that the characteristic polynomial reflects many properties of the corresponding linear operator.

Bearing these facts in mind, it is not surprising that the characteristic polynomial of the adjacency matrix of a graph and its properties have been widely studied. The first papers dealing with investigations of this kind were published in 1956-1957 by the Soviet mathematician Lichtenbaum [15] and by two German mathematicians – Collatz and Sinogowits [16]. (It is worth noting here that the main results of ref. 16 were obtained much earlier – during the Second World War.) The ideas of Lichtenbaum and Collatz and Sinogowitz proved to be very fruitful, and their work soon developed into so-called graph spectral theory [1]. Many hundreds of papers have since been published in this area.

4.4.1.1. DEFINITION. *If* A *is the adjacency matrix of a graph* G *and* I *is the unit matrix of order* n, *the characteristic polynomial of the graph* G *is defined as:*

$$\mathrm{Ch}(G, x) = \det(x\mathbf{I} - \mathbf{A}).$$

In other words, the characteristic polynomial of a graph is just the characteristic polynomial of its adjacency matrix. When there is no possibility of any misunderstanding, we shall denote the characteristic polynomial, defined in 4.4.1.1, as $\mathrm{Ch}(G)$ or $\mathrm{Ch}(x)$ or simply as Ch.

$\mathrm{Ch}(x)$ is a polynomial of degree n and will be presented in the form:

$$\mathrm{Ch}(G, x) = x^n + c(G, 1)x^{n-1} + c(G, 2)x^{n-2} + \ldots$$

$$+ c(G, \ n-1)x + c(G, n),$$

or, using shorthand notations, as:

$$\mathrm{Ch}(G, x) = \sum_{k=0}^{n} c(G, k)x^{n-k},$$

where $c(G, 0) = 1$ for all graphs G. When misunderstanding is not possible, we shall write $c(G, k) = c(k)$.

Two elementary but essential properties of $\mathrm{Ch}(G, x)$ will be given first.

4.4.1.2. THEOREM. *If the graphs* G_a *and* G_b *are isomorphic, then their characteristic polynomials coincide, i.e.* $\mathrm{Ch}(G_a) = \mathrm{Ch}(G_b)$.

This is equivalent to stating that the characteristic polynomial is a graph invariant which is independent of the labeling of the vertices. (Note that the adjacency matrix does depend on the labeling of the vertices of the corresponding graph. Two isomorphic graphs need not have equal adjacency matrices.)

4.4.1.3. THEOREM. *The converse of Theorem 4.4.1.2 is, unfortunately, not true. Thus, if G_a and G_b are two graphs having equal characteristic polynomials, $\mathrm{Ch}(G_a) = \mathrm{Ch}(G_b)$, it is not necessarily true that G_a and G_b are isomorphic.*

If two nonisomorphic graphs G_a and G_b have equal characteristic polynomials, we say that G_a and G_b are isospectral. Examples of isospectral graphs are well known (and were first observed by Collatz and Sinogowitz [16]). The smallest two such graphs are G_{51} and G_{52}:

It is also not difficult to construct isospectral molecular graphs. Two examples are the graphs G_{53} and G_{54} (the molecular graphs of 2,3-dimethylheptane and 2-methyl, 4-ethylhexane, respectively); and the graphs G_{55} and G_{56} (the molecular graphs of p-divinylbenzene and 2-phenylbuitadiene, respectively).

There is currently an extensive mathematical and chemical literature on the problem of designing isospectral graphs. We shall not go into detail here; the interested should consult the references cited [1,17,182]. Some of the most impressive results in this area were discovered by Schwenk [18].

4.4.1.4. THEOREM (SCHWENK). *For arbitrarily large values of k, there exist collections of k mutually nonisomorphic but mutually isospectral graphs.*

4.4.1.5. THEOREM (SCHWENK). *If T is a randomly chosen tree with n vertices, the probability that T has an isospectral pair tends to unity when n tends to infinity.*

This means that almost all trees have isospectral partners or, in other words, that almost all trees are *not* determined by their characteristic polynomials. It is likely that Schwenk's Theorem 4.4.1.5 can be extended to cyclic graphs as well, though this remains still to be proved.

In the formulation of Theorem 4.4.1.3 we have used the word "unfortunately." This is because the existence of isospectral graphs indicates that a substantial part of the information about the structure of a graph G has been lost in $Ch(G)$. The theory of the characteristic polynomial is thus less rich than the theory of graphs and many problems in graph theory are simply intractable by means of graph spectral theory. As a consequence, the power of polynomial techniques in graph theory is much reduced. Similarly, several difficulties occur in the applications of Ch in chemistry.

For example, isospectral molecular graphs should correspond to conjugated molecules having identical or (at least) closely similar photoelectron spectra, ionization potentials, etc. No analogous behavior has been observed experimentally [19].

The idea of using $Ch(G)$ in chemical documentation work for the coding of molecular structure had to be abandoned because of the occurrence of isospectral molecular graphs [20].

In spite of the pessimistic outlook the reader may have after reading Theorems 4.4.1.3 to 4.4.1.5, there are several structural features of a graph which can be reconstructed from its characteristic polynomial. In the following we list some elementary examples [1].

4.4.1.6. THEOREM. *For a simple graph G*

$$
\begin{aligned}
c(1) &= 0 \\
-c(2) &= m = \text{number of edges of G} \\
-c(3)/2 &= n_3 = \text{number of triangles in G}.
\end{aligned}
$$

4.4.1.7. THEOREM. *If g is the size of the smallest odd cycle in the graph G, then $c(g - 2k) = 0$ for $k = 1, 2, \ldots, (g - 1)/2$ and $c(g) \neq 0$. Furthermore, $-c(g)/2$ is the number of cycles of size g in G.*

4.4.1.8. THEOREM. *A graph is bipartite if and only if $c(2k + 1) = 0$ for all k. If a graph is bipartite, we have $(-1)^k c(2k) \geq 0$ for all k.*

4.4.1.9. THEOREM. *A graph* G *is acyclic if and only if* $c(2k+1) = 0$ *for all* k, *and* $c(2k) = (-1)^k m(G,k)$, *where* $m(G,k)$ *is the number of* k-*matchings of* G *(see later).*

To conclude this section we shall calculate, as an example, the characteristic polynomial of the four-membered cycle C_4:

$$A(C_4) = \begin{bmatrix} 0 & 1 & 0 & 1 \\ 1 & 0 & 1 & 0 \\ 0 & 1 & 0 & 1 \\ 1 & 0 & 1 & 0 \end{bmatrix}.$$

By applying Definition 4.4.1.1, we obtain:

$$\mathrm{Ch}(C_4, x) = \det \begin{bmatrix} x & -1 & 0 & -1 \\ -1 & x & -1 & 0 \\ 0 & -1 & x & -1 \\ -1 & 0 & -1 & x \end{bmatrix}.$$

Expanding this determinant according to standard rules we find that

$$\mathrm{Ch}(C_4, x) = x \begin{bmatrix} x & -1 & 0 \\ -1 & x & -1 \\ 0 & -1 & x \end{bmatrix} - (-1) \begin{bmatrix} -1 & -1 & 0 \\ 0 & x & -1 \\ -1 & -1 & x \end{bmatrix} +$$

$$+ 0 \cdot \begin{bmatrix} -1 & x & 0 \\ 0 & -1 & -1 \\ -1 & 0 & x \end{bmatrix} - (-1) \begin{bmatrix} -1 & x & -1 \\ 0 & -1 & x \\ -1 & 0 & -1 \end{bmatrix}$$

$$= x(x^3 - 2x) + (-x^2 - 1 + 1) + (-1 - x^2 + 1)$$

and finally that

$$\mathrm{Ch}(C_4, x) = x^4 - 4x^2.$$

4.4.2 Some Comments of Graph Spectral Theory

4.4.2.1. DEFINITION. *If* A *is the adjacency matrix of a graph* G, *and if there is a non-zero vector* C *and a scalar* x *such that:*

$$AC = xC$$

then C *is called an eigenvector of the graph G and x is an eigenvalue of the graph G. The eigenvalue x corresponds to the eigenvector C.*

4.4.2.2. DEFINITION. *If the graph G has q linearly independent eigenvectors with the same eigenvalue x, we say x is an eigenvalue of G with multiplicity q.*

4.4.2.3. DEFINITION. *The collection of all eigenvalues of a graph (taking into account their multiplicities) is called the spectrum of the graph.*

The spectrum of a graph is of some importance in theoretical chemistry because of the intimate relationship between the eigenvalues of the molecular graph and the Hückel molecular orbital energy levels. In this section we shall briefly summarize the basic facts concerning the spectrum of a graph and its relation to the characteristic polynomial. As already mentioned, the theory of graph spectra is treated fully in the monograph cited in ref. 1.

4.4.2.4. THEOREM. *For a graph G having n vertices, there will be n graph eigenvalues. If these eigenvalues are denoted by $x_1, x_2, \ldots, x_n$, we may write:*

$$(x - x_1)(x - x_2)\ldots(x - x_n) = \mathrm{Ch}(G, x).$$

This is equivalent to stating that the graph eigenvalues are the roots of the equation:

$$\mathrm{Ch}(x) = 0.$$

The spectrum of the four-membered cycle C_4 consists, for instance, of the numbers $+2$, 0, 0 and -2 because the roots of the equation $\mathrm{Ch}(C_4) = 0$, i.e., $x^4 - 4x^2 = 0$ are clearly the above listed numbers.

4.4.2.5. THEOREM. *The eigenvalues of a simple graph are real numbers. This implies that all the zeros of the characteristic polynomial of such a graph are real.*

Note that Theorem 4.4.2.5 is an elementary consequence of the fact that the eigenvalues of a symmetric matrix are real.

If the graph G contains directed edges (i.e., if G is a digraph and $\mathbf{A}(G)$ is not symmetric), its eigenvalues need not all be real. The conditions under which a digraph has a real spectrum are not known.

According to Theorem 4.4.2.1, the spectrum of a graph is fully determined by the characteristic polynomial and *vice versa* [21]. There is thus no essential difference between graph spectral theory and the theory of the characteristic polynomial. The two theories are completely equivalent. Both the graph spectrum and the characteristic polynomial contain exactly the same information about the structure of the graph. In some cases, it is more natural to use the language of graph spectral theory whereas in other cases the polynomial formalism is more appropriate. To exemplify this we present the spectral analogy of Theorem 4.4.1.8, which is the basis of the famous Pairing theorem [22].

4.4.2.6. THEOREM. *A graph is bipartite if and only if* $x_i + x_{n+1-i} = 0$ *for all* $i = 1, 2, \ldots, n$. *(Every eigenvalue* x_i *thus has a "pair", namely* x_{n+1-i}.*)*

As a direct consequence of Theorem 4.4.2.6, a bipartite graph with an odd number of vertices necessarily possesses a zero in its spectrum.

An improved spectral characterization of bipartite graphs was discovered by Cvetković [23].

4.4.2.7. THEOREM. *Let* x_1 *denote the greatest and* x_n *the smallest eigenvalue of a graph. A connected graph is bipartite if and only if* $x_1 + x_n = 0$.

4.4.3 The Theorem of Sachs

One of the most fundamental questions that can be posed in connection with any graphic polynomial concerns the relation between the structure of a graph and the coefficients of the respective polynomial. In the case of the characteristic polynomial, the solution was obtained by the German mathematician Horst Sachs [24] in 1963. Before stating the theorem of Sachs, we would like to point out that a number of other researchers independently and almost simultaneously reached the same result. The work of Harary [25], Milić [26] and Spialter [27] should be especially mentioned here. A full bibliography of the history of the discovery of Sachs' theorem can be found elsewhere [1].

4.4.3.1. DEFINITION. *A graph all of whose components are isomorphic to* K_2 *or* C_3 *or* C_4 *or* C_5 *or* C_6 *etc. is called a Sachs graph. A Sachs graph* s *has* $n(s)$ *vertices and* $p(s)$ *components,* $c(s)$ *of which are cycles.*

(In the original paper of Sachs [24] the name "basic graph" was used. The names "characteristic graph" and "mutation graph" were also proposed for the same notion. In the present chapter the choice of "Sachs graph" follows the terminology of the earliest publications on the chemical application of Sachs' theorem [28–32].)

For the Sachs' graphs s_1, s_2 and s_3 illustrated overleaf:

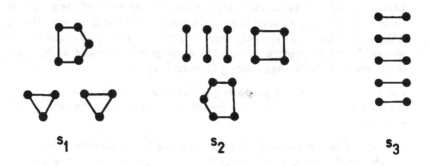

$$s_1 \qquad\qquad\qquad\qquad s_2 \qquad\qquad\qquad\qquad s_3$$

The reader is invited to check that:

$$n(s_1) = 11, \quad p(s_1) = 3, \quad c(s_1) = 3$$

$$n(s_2) = 15, \quad p(s_2) = 5, \quad c(s_2) = 2$$

$$n(s_3) = 10, \quad p(s_3) = 5, \quad c(s_3) = 0.$$

4.4.3.2.   THEOREM OF SACHS.   *If $c(G,k)$ is the coefficient of the characteristic polynomial, then*

$$c(G,k) = \sum_s (-1)^{p(S)} 2^{c(s)}$$

with the summation extending over all the Sachs' graphs s with k vertices contained as subgraphs in the graph G ($k > 0$). If in the graph G there are no Sachs' graphs with the required number of vertices, then $c(G,k) = 0$.

Another formulation of the Sachs' theorem is the following.

4.4.3.3. THEOREM.

$$\text{Ch}(G,x) = x^n + \sum_s (-1)^{P}(s) 2^{c(s)} x^{n-n(s)}$$

where the summation extends over all Sachs' graphs s contained as subgraphs in the graph G.

Because of its important role in theoretical chemistry, Sachs' theorem has been stated and exemplified in a large number of chemical publications

[2–4, 9, 28, 31, 33–36]. Here we shall exemplify the use of Sachs' theorem by considering the graph C_4. This graph contains seven Sachs' graphs, i.e., $s_1 - s_7$. Using

$$C_4 \qquad s_1 \qquad s_2 \qquad s_3 \qquad s_4 \qquad s_5 \qquad s_6 \qquad s_7$$

Theorem 4.4.3.3 we obtain

$$\mathrm{Ch}(C_4, x) = x^4 + \sum_{i=1}^{7} (-1)^{p(S_i)} 2^{c(s_i)} x^{4-n(s_i)}.$$

Now

$$n(s_1) = n(s_2) = n(s_3) = n(s_4) = 2, \qquad n(s_5) = n(s_6) = n(s_7) = 4$$

$$p(s_1) = p(s_2) = p(s_3) = p(s_4) = p(s_7) = 1, \qquad p(s_5) = p(s_6) = 2$$

$$c(s_1) = c(s_2) = c(s_3) = c(s_4) = c(s_5) = c(s_6) = 0, \qquad c(s_7) = 1,$$

and therefore

$$\mathrm{Ch}(C_4, x) = x^4 + [(-1)^1 2^0 x^{4-2}] \cdot 4 + [(-1)^2 2^0 x^{4-4}] \cdot 2$$
$$+ [(-1)^1 2^1 x^{4-4}] = x^4 - 4x^2.$$

It would be useful for the reader to convince himself that:

$$\mathrm{Ch}(G_{57}) = x^4 - 3x^2 + 1$$
$$\mathrm{Ch}(G_{58}) = x^4 - 4x^2 - 2x + 1$$
$$\mathrm{Ch}(G_{59}) = x^6 - 6x^4 + 9x^2 - 4.$$

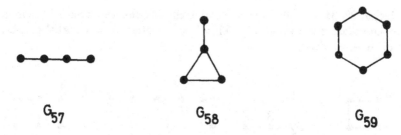

G_{57} G_{58} G_{59}

After performing such exercises, verification of the statements in Theorem 4.4.1.6 will be quite elementary. As a more advanced application of Sachs' theorem we now prove the previously stated Theorem 4.4.1.7.

4.4.3.4. PROOF OF 4.4.1.7. Note first that a Şachs' graph with an odd number of vertices necessarily contains an odd-membered cycle. Now the size of the smallest odd cycle in G is g and there are therefore no Sachs' graphs in G with $g - 2$, $g - 4$ etc. vertices. Consequently, $c(g - 2) = c(g - 4) = \cdots = 0$. Every Sachs' graph on g vertices must contain a cycle with g vertices and therefore no other component: $n(s) = g$, $p(s) = 1$, $c(s) = 1$. The number of such Sachs' graph is obviously equal to the number n_g of cycles of size g in the graph G. From Sachs' theorem it follows that:

$$c(g) = n_g(-1)^1 2^1 = -2n_g.$$

This completes the proof of Theorem 4.4.1.7.

4.4.4 Relations Between Characteristic Polynomials

Knowledge of the relations between characteristic polynomials is very important because such relations enable one to deduce the properties of complex graphs from those of relatively simple graphs. Several of the results presented in this section are also useful for paper-and-pencil calculations of Ch.

The following two relations are elementary consequences of Definition 4.4.1.1 and well-known properties of determinants.

4.4.4.1. THEOREM. *If the graph G is composed of two disconnected components G_a and G_b, then $\mathrm{Ch}(G) = \mathrm{Ch}(G_a) \cdot \mathrm{Ch}(G_b)$. In the general case: if the graph G is composed of components $G_a \cdot G_b, \ldots, G_p$, then $\mathrm{Ch}(G) = \mathrm{Ch}(G_a) \cdot \mathrm{Ch}(G_b) \ldots \mathrm{Ch}(G_p)$.*

4.4.4.2. Theorem. *For the vertices* $v_1, v_2, \ldots, v_n$ *of* G, *it can be proven that*

$$\frac{d\text{Ch}(G, x)}{dx} = \sum_{i=1}^{n} \text{Ch}(G - v_i, x).$$

Whereas the determinantal equivalent of Theorem 4.4.4.2 has been known for over a century, its graph-theoretic counterpart first seems to have been used by Clarke [37]. In the chemical literature there has been some confusion on the interpretation of this result [38], a matter which has recently been resolved [39]. It is also worthy of mention here that Randić [40] developed a novel technique for the computation of Ch(G) based on Theorem 4.4.4.2.

The basic recurrence relation for Ch(G) seems to have been discovered first by Heilbronner [41] in 1953. Schwenk reached the same formula some twenty years later [42]. In the meantime, Hosoya had also formulated results equivalent to those of Heilbronner [43,44].

4.4.4.3. Theorem (Heilbronner). *If* v_r *and* v_s *are two vertices of the graph* G, *joined by the edge* e_{rs}, *then*

$$\text{Ch}(G) = \text{Ch}(G - e_{rs}) - \text{Ch}(G - v_r - v_s) - 2 \sum_{g} \text{Ch}(G - Z)$$

where the summation extends over all cycles of the graph G *containing the edge* e_{rs}.

Theorem 4.4.4.3 has several important special cases.

4.4.4.4. Corollary. *If the edge* e_{rs} *is a bridge, i.e.,* e_{rs} *does not belong to any cycle, then*

$$\text{Ch}(G) = \text{Ch}(G - e_{rs}) - \text{Ch}(G - v_r - v_s).$$

In addition, the graphs $G - e_{rs}$ and $G - v_r - v_s$ are necessarily disconnected, and thus the above equation can be further simplified by applying Theorem 4.4.4.1. An interesting application of 4.4.4.4 is the "Polynomial matrix method" of Kaulgud and Chitgopkar [45–47]. For a generalization of 4.4.4.4 see ref. 48.

The only case in which the subgraph $G - v_r - v_s$ is connected occurs when v_r (or v_s) has degree one.

4.4.4.5. COROLLARY. *When vertex v_r has degree one, we obtain:*

$$Ch(G) = x\,Ch(G - v_r) - Ch(G - v_r - v_s).$$

4.4.4.6. COROLLARY. *If P_n is the path on n vertices, we obtain:*

$$Ch(P_n) = x\,Ch(P_{n-1}) - Ch(P_{n-2}).$$

Since $Ch(P_1) = x$ and $Ch(P_2) = x^2 - 1$, by making use of 4.4.4.6 we can easily compute the characteristic polynomials of P_n for larger values of n. As examples, we present the following:

$$Ch(P_1) = x$$
$$Ch(P_2) = x^2 - 1$$
$$Ch(P_3) = x^3 - 2x$$
$$Ch(P_4) = x^4 - 3x^2 + 1$$
$$Ch(P_5) = x^5 - 4x^3 + 3x$$
$$Ch(P_6) = x^6 - 5x^4 + 6x^2 - 1$$
$$Ch(P_7) = x^7 - 6x^5 + 10x^3 - 4x$$
$$Ch(P_8) = x^8 - 7x^6 + 15x^4 - 10x^2 + 1$$
$$Ch(P_9) = x^9 - 8x^7 + 21x^5 - 20x^3 + 5x$$
$$Ch(P_{10}) = x^{10} - 9x^8 + 28x^6 - 35x^4 + 15x^2 - 1.$$

Further generalizations and corollaries of Heilbronner's result 4.4.4.3, and some additional recursion relations for $Ch(x)$ can be found elsewhere [40, 49–55].

4.4.4.7. THEOREM. *Let P_{rs} be a path in the graph G, connecting the vertices v_r and v_s. It follows that:*

$$Ch(G - v_r)Ch(G - v_s) - Ch(G)Ch(G - v_r - v_s) = \left[\sum Ch(G - P_{rs}) \right]^2,$$

with the summation on the r.h.s. ranging over all paths in G connecting v_r with v_s.

The above result was known to Coulson and Longuet-Higgins [56] though it was later overlooked in both graph spectral theory and the topological theory of conjugated molecules. Recently, Theorem 4.4.4.7 was used to prove the Law of alternating polarity [57].

As an illustration of the applicability of the above relations we now calculate $Ch(C_4)$. From 4.4.4.3,

$$Ch\left(\;\square\;\right) = Ch\left(\;\llcorner\;\right) - Ch\left(\;\mid\;\right) - 2\,Ch(\emptyset)$$

where $\emptyset$ symbolizes the graph have no vertices. Since $Ch(\emptyset) = 1$ by definition, we have shown that

$$Ch(C_4) = Ch(P_4) - Ch(P_2) - 2.$$

Using the above listed expressions for $Ch(P_n)$, we obtain immediately that $Ch(C_4) = (x^4 - 3x^2 + 1) - (x^2 - 1) - 2 = x^4 - 4x^2$, in full agreement with the previous results.

Note that in the general case this equation becomes:

$$Ch(C_n) = Ch(P_n) - Ch(P_{n-2}) - 2.$$

As another, more advanced illustration, we determine next the Ch of the molecular graph of naphthalene. By Theorem 4.4.4.3,

$$Ch\left(\;\bigcirc\!\bigcirc\;\right) = Ch\left(\;\bigcirc\!\bigcirc\;\right) - Ch\left(\;\bigcirc\;\bigcirc\;\right)$$

$$- 2\cdot Ch\left(\;\bigcirc\;\right) - 2\cdot Ch\left(\;\bigcirc\;\right)$$

and

$$Ch\left(\;\bigcirc\!\bigcirc\;\right) = Ch\left(\;\bigcirc\!\bigcirc\;\right) - Ch\left(\;\bigcirc\!\bigcirc\;\right) - 2\cdot Ch(\emptyset)$$

Using Theorem 4.4.4.1 we find further that:

$$\text{Ch}(\;\text{⬡⬡}\;) = \text{Ch}(P_{10}) - \text{Ch}(P_8) - \text{Ch}(P_4)^2 - 4\;\text{Ch}(P_4) - 2$$

Substituting in the above formula the expressions for $CH(P_n)$ previously obtained, one arrives to the final result

$$\text{Ch}(\;\text{⬡⬡}\;) = x^{10} - 11x^8 + 41x^6 - 65x^4 + 43x^2 - 9$$

Details of the extensive recent work on characteristic polynomials can be found elsewhere [157,183–185]. In a series of papers [186–189] Dias examined some structural invariants of the molecular graph, related to the coefficients of the characteristic polynomial and having chemical significance. Various methods for the computation of the characteristic polynomials of chemical graphs have been put forward [190–194]. The factorization of the characteristic polynomials was studied by Kirby [195]. In the present moment (January 1989) it seems that the characteristic polynomial still remains the most popular among graphic polynomials of mathematical chemistry.

4.4.5   Chemical Applications of Characteristic Polynomials

The characteristic polynomial of the molecular graph and its zeros (i.e., the graph spectrum) are of considerable importance in theoretical chemistry. Their applications in Hückel molecular orbital theory are treated in detail in another chapter of this book. Attempts to use the coefficients of $Cg(G)$ in chemical documentation [27,58] were unsuccessful [20] because of the existence of many isospectral graphs. The largest zero of $\text{Ch}(G)$ has been proposed as a measure of the extent of branching in the corresponding molecule [59–61].

4.5   The Matching Polynomials

4.5.1   Introduction to Matching Polynomials

Two edges in a graph are said to be independent if they have no common vertex. For example, the edges e_{12} and e_{34} or the edges e_{25} and e_{45} in the graph G_{61} are independent.

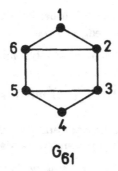

G_{61}

The edges e_{12} and e_{26} of G_{61} are not independent, however, since they have a common vertex (v_2).

4.5.1.1. DEFINITION. *A collection of k mutually (i.e., pairwise) independent edges in a graph $G(k \geq 2)$ is called a k-matching of G.*

If $k = n/2$, the corresponding matching is described as perfect. Perfect matchings play a major role in graph theory. They are in a one-to-one correspondence with the Kekulé structural formulas of conjugated molecules and are thus of considerable importance in chemical applications of graph theory too. As this topic will not be discussed here, the interested reader should consult the reviews [62,63].

4.5.1.2. DEFINITION. *The number of k-matchings of the graph G will be denoted as $m(G,k)$. In addition, $m(G,1) = m =$ number of edges in the graph G, and $m(G,O) = 1$ for all graphs G.*

Note that if the graph G represents a conjugated molecule with n conjugated centers ($n =$ even), $m(G,n/2)$ is equal to the number of Kekulé structural formulas of the respective molecule. This is a quantity of crucial importance in resonance theory [62].

Below we determine the numbers $m(G,k)$ of the graph G_{61}. We obtain:

$m(G_{61},0) = 1$ by definition

$m(G_{61},1) = 8$ equals number of edges

$m(G_{61},2) = 14$ because the following fourteen pairs of edges are independent: (e_{12},e_{34}), (e_{12},e_{35}), (e_{12},e_{45}), (e_{12},e_{56}), (e_{16},e_{23}), (e_{16},e_{34}), (e_{16},e_{35}), (e_{16},e_{45}), (e_{23},e_{45}), (e_{23},e_{56}), (e_{26},e_{34}), (e_{26},e_{35}), (e_{26},e_{45}), and (e_{34},e_{56})

$m(G_{61},3) = 2$ because the following two triplets of edges are mutually independent: (e_{12},e_{34},e_{56}) and (e_{23},e_{45},e_{61})

$m(G_{61}, 4) = m(G_{61}, 5) =$ etc. $= 0$ because it is not possible to select more than three independent edges in G_{61}.

4.5.1.3. DEFINITION. *The matching polynomial of a graph G is defined as:*

$$Ma(G, x) = \sum_{k=0}^{m} (-1)^k m(G, k) x^{n-2k}.$$

When no misunderstanding can arise, we shall denote the matching polynomial as $Ma(G)$ or $Ma(x)$ or simply as Ma.

4.5.1.4. DEFINITION. *The generating function of the number of matchings in a graph G is given by:*

$$M(G, x) = \sum_{k=0}^{m} m(G, k) x^k.$$

Of course, there is a one-to-one correspondence between the polynomials $Ma(G, x)$ and $M(G \ x)$. Obtaining this relation may serve as an elementary exercise to the interested reader.

For example, for the above-examined graph G_{61}

$$Ma(G_{61}) = x^6 - 8x^4 + 14x^2 - 2$$

$$M(G_{61}) = 1 + 8x + 14x^2 + 2x^3.$$

Although the matching polynomial has been introduced in chemistry, physics and mathematics only relatively recently, its history is quite unusual. The mere fact that in the period of only ten years Ma was independently discovered at least six times is a convincing argument in favour of its wide importance.

In statistical physics the concept of matching appears in a natural way in the study of the absorption of diatonic molecules (in physical terminology: dimers) on a crystal lattice. In 1970 Heilmann and Lieb [64], and independently Kunz [65], considered problems in statistical physics in which the matching polynomial of a graph was related to the partition function of a so-called "monomer-dimer" system. In order to be able to describe a phase transition in such a system, it was necessary that the matching polynomial possess complex zeros. However, in both references [64] and [65] it was demonstrated that all the zeros of the matching polynomial are real numbers. This finding was rather inconvenient for the monomer-dimer theory, since it implied that in terms of the model considered, phase transitions cannot be described. The theory was later extensively elaborated and, consequently, numerous properties of Ma were determined [66,67].

The next discoverer of the matching polynomial was Hosoya [43,44], who considered the polynomial $M(G,x)$ of the molecular graph of a saturated hydrocarbon. In particular, Hosoya noticed that the total number of matchings in G, namely

$$Z(G) = M(G,1) = \sum_{k=0}^{m} m(G,k),$$

can be correlated with various physicochemical properties of the corresponding hydrocarbon (boiling point, etc.). $Z(G)$ is nowadays usually named the topological index of Hosoya. It is, in fact, a measure of the extent of branching of the carbon atom skeleton of a hydrocarbon. A detailed exposition of the theory of the Hosoya index can be found in Chapter 11 of the book [162]. For recent investigations of the mathematical properties of $Z(G)$ see [163-165,196].

The matching polynomial Ma(G,x) was introduced by Gutman *et al* [68-71] in connection with the theory of the so-called topological resonance energy (see later). In these papers Ma(G,x) was referred to as the "acyclic polynomial". Independently, Aihara [72] developed a fully equivalent resonance energy concept; he named Ma(G,x) the "reference polynomial".

Obviously unaware of all this previous work, Farrell published in 1979 a mathematical paper [73], introducing the matching polynomial for the sixth time. In fact, there is a slight and inessential difference [74] between Farrell's "matching polynomial" and the matching polynomial defined in the present work.

In concluding this short history, it should also be mentioned that there is a close relation between the so-called rook polynomial [75] and the matching polynomial. Every rook polynomial coincides with the matching polynomial of a bipartite graph [76].

The theory of the matching polynomial has been expounded in greater detail in [77, 78, 157]. Methods for computation of the matching polynomial are outlined in [197].

We present first a few elementary properties of Ma(G).

4.5.1.5. THEOREM. *If two graphs G_a and G_b are isomorphic, their matching polynomials coincide, Ma(G_a) = Ma(G_b).*

4.5.1.6. THEOREM. *The reverse of Theorem 4.5.1.5 is not true. There exist pairs of nonisomorphic graphs G_a and G_b, such that $Ma(G_a) = Ma(G_b)$.*

It is very easy indeed to find such pairs. The smallest graphs of this kind are G_{62} and G_{63}. The smallest pair of connected graphs with equal Ma are G_{64} and G_{65}.

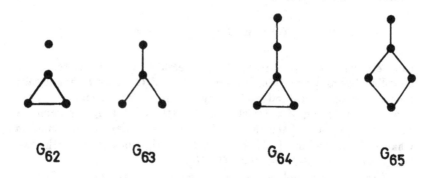

$$G_{62} \qquad G_{63} \qquad G_{64} \qquad G_{65}$$

The consequence of Theorem 4.5.1.6 are analogous to those of Theorem 4.4.1.3: $Ma(G)$ incorporates only a small part of the information on the structure of the graph G. From the above examples is seen that from $Ma(G)$ one cannot deduce even whether G is connected or bipartite.

4.5.1.7. THEOREM. *If the graph G is composed of two disconnected components G_a and G_b, then $Ma(G) = Ma(G_a) \cdot Ma(G_b)$. In the general case: if G is composed of components $G_a, G_b, \ldots, G_p$, then $Ma(G) = Ma(G_a) \cdot Ma(G_b) \ldots Ma(G_p)$.*

4.5.1.8. THEOREM. *If e_{rs} is an (arbitrary) edge of the graph G connecting the vertices v_r and v_s, it follows that:*

$$Ma(G) = Ma(G - e_{rs}) - Ma(G - v_r - v_s).$$

Because of the great importance of Theorem 4.5.1.8 for the calculation of Ma, we present its proof [69,71,73] below:

4.5.1.9. PROOF OF 4.5.1.8. Consider the $m(G,k)$ selections of k independent edges (i.e., k-matchings) in the graph G. These k-matchings can be divided into two groups. In the first group are the k-matching of G which do not contain the edge e_{rs}, and there are evidently $m(G - e_{rs}, k)$ such selections. In the second group are the k-matchings which contain the edge e_{rs}. In order to obtain such a k-matching, we have to select e_{rs} and an additional $k - 1$ independent edges. These latter, however, must not be incident to the vertices v_r and v_s. Consequently, we have to select $k - 1$ independent edges from $G - v_r - v_s$, and the number of k-matchings of G containing the edge e_{rs} is $m(G - v_r - v_s, k - 1)$. We thus obtain the result:

$$m(G, k) = m(G - e_{rs}, k) + m(G - v_r - v_s, k - 1).$$

Theorem 4.5.1.8 can now be derived by combining the above relation with Definition 4.5.1.3.

Theorem 4.5.1.8 has a number of important corollaries.

4.5.1.10. COROLLARY.

$$Ma(P_n) = xMa(P_{n-1}) - Ma(P_{n-1}).$$

In addition, $Ma(P_1) = x$ and $Ma(P_2) = x^2 - 1$ (the reader should check this!). Comparing this result with Corollary 4.4.4.6, we see that the matching and the characteristic polynomials of the path P_n coincide. Hence we may use the previously tabulated expressions for $Ch(P_n)$ in the theory of the matching polynomial. We shall see later that the matching and characteristic polynomials coincide in the case of all acyclic graphs.

4.5.1.11. COROLLARY.

$$Ma(C_n) = Ma(P_n) - Ma(P_{n-2})$$

$$Ma(C_n) = xMa(C_{n-1}) - Ma(C_{n-2}).$$

4.5.1.12. COROLLARY. *If the vertex v of the graph G is adjacent to the vertices $w_1, w_2, \ldots, w_d$, then*

$$Ma(G) = xMa(G - v) - \sum_{i=1}^{d} Ma(G - v - v_i).$$

4.5.1.13. PROOF OF 4.5.1.12. Apply Theorem 4.5.1.8 successively to the edges connecting v and w_i, with $i = 1, 2, \ldots, d$, and note that by deleting

of all these edges one obtains a graph whose components are $G - v$ and an isolated vertex. Since the matching polynomial of an isolated vertex is just x, Corollary 4.5.1.12 follows from Theorem 4.5.1.7.

Using the result 4.5.1.12 we can derive a sample recurrence relation for the matching polynomial of the complete graph K_n. Since every vertex v of K_n has exactly $n - 1$ neighbors $(w_1, w_2, \ldots, w_{n-1},$ say$)$, and since $K_n - v = K_{n-1}$ and $K_n - v - w_i = K_{n-2}$, we can immediately conclude that:

$$\mathrm{Ma}(K_n) = x\mathrm{Ma}(K_{n-1}) - (n - 1)\mathrm{Ma}(K_{n-2}).$$

It is also immediately evident that:

$$\mathrm{Ma}(K_1) = x \quad \text{and} \quad \mathrm{Ma}(K_2) = x^2 - 1.$$

Bearing in mind that the Hermite polynomials $H_n(x)$ obey the recursion relation:

$$H_n(x) = xH_{n-1}(x) = (n - 1)H_{n-2}(x),$$

and that $H_1(x) = x$, $H_2(x) = x^2 - 1$, we arrive at the unexpected result that Hermite polynomials fully coincide with the matching polynomials of complete graphs.

There are, in fact, several such relations [79].

4.5.1.14. THEOREM. *Let $T_n(x)$ and $U_n(x)$ be the Chebyshev polynomial of the first kind and the Chebyshev function of the second kind, respectively. Let $H_n(x)$ be the Hermite polynomial. Then*

$$T_n(x) = \frac{1}{2}\mathrm{Ma}(C_n, 2x) \qquad n \geq 3$$

$$U_n(x) = \sqrt{1 - x^2}\,\mathrm{Ma}(P_{n-1}, 2x) \qquad n \geq 2$$

$$H_n(x) = \mathrm{Ma}(K_n, x) \qquad n \geq 1.$$

Similar results are also known for the Laguerre and Legendre polynomials [79,80].

Using the above theorem and the well-known analytical properties of Hermite polynomials, Godsil [81] was able to prove the following relation.

4.5.1.15. COROLLARY. *If G is a graph and $\overline{G}$ is its complement, then*

$$\frac{1}{\sqrt{2\pi}} \int_{-\infty}^{+\infty} \mathrm{Ma}(G, x)e^{-x^2/2}dx$$

is equal to the number of perfect matchings of $\overline{G}$.

We mention here in passing that the matching polynomial of $\overline{G}$ can be computed from the matching polynomial G [81,82].

4.5.1.16. THEOREM.

$$\text{Ma}(\overline{G}, x) = \sum_{k=0}^{m} m(G, k)\text{Ma}(K_{n-2k}, x).$$

Theorems 4.5.1.7 and 4.5.1.8 provide an easy and efficient recursion technique for the computation of the matching polynomial of a molecular graph. In such computations it is very useful to take into account the results of Corollary 4.5.1.10.

By way of illustration we determine $\text{Ma}(G_{61})$. Upon applying Theorem 4.5.1.8 we obtain:

Combining these relations and using Theorem 4.5.1.7 we find that:

$$\text{Ma}(G_{61}) = \text{Ma}(C_3)^2 - \text{Ma}(P_2)^2 - \text{Ma}(P_4).$$

From 4.5.1.11 it can be immediately verified that $\text{Ma}(C_3) = x^3 - 3x$ and then using the tables of $\text{Ch}(P_n)$ we obtain the result:

$$\text{Ma}(G_{61}) = (x^3 - 3x)^2 - (x^2 - 1)^2 - x^4 - 3x^2 + 1) =$$

$$= x^6 - 8x^4 + 14x^2 - 2,$$

in full agreement with the previous direct calculation.

The reader may like to verify that

$$\text{Ma}(G_{58}) = x^4 - 4x^2 + 1$$

$$\text{Ma}(G_{59}) = x^6 - 6x^4 + 9x^2 - 2,$$

where G_{58} and G_{59} are graphs given in the previous section. $Ma(G_{58})$ and $Ma(G_{59})$ may be computed in two ways: by direct enumeration of all matchings (Definition 4.5.1.3) and by using Theorem 4.5.1.8.

4.5.2. Relations Between the Matching and the Characteristic Polynomial

If one compares the definition of the characteristic polynomial (4.4.1.1) with that of the matching polynomial (4.5.1.3), the wrong impression may be obtained that these two graphic polynomials have hardly any common feature. That this is not so will be demonstrated in the present section. The relations between $Ma(G)$ and $Ch(G)$ represent an important and nontrivial part of the theory of matching polynomials and are of the greatest value in chemical applications.

Combining Definition 4.5.1.3 with Theorem 4.4.1.9, we reach the following immediate yet significant result.

4.5.2.1. THEOREM. *The matching and the characteristic polynomials of a graph G coincide, i.e., $Ma(G) = Ch(G)$, if and only if G is a forest.*

4.5.2.2. PROOF OF 4.5.2.1 AND 4.4.1.9. Suppose G is an acyclic graph and apply Sachs' theorem 4.4.3.2. Since G contains no odd cycles, we may immediately conclude that G contains no Sachs' graphs with an odd number of vertices. Therefore

$$c(k) = 0 \text{ if } k \text{ is odd.}$$

If k is even, G may contain Sachs' graphs having k vertices. All these Sachs' graphs must be composed of $k/2$ components K_2 (since G is assumed to be acyclic). In other words, $p(s) = k/2$ and $c(s) = 0$ for all the Sachs' graphs of G. Every such Sachs' graph corresponds in an obvious manner to a selection of $k/2$ independent edges in G. Therefore

$$c(k) = m(G, k/2) \cdot (-1)^{k/2} 2^0 = (-1)^{k/2} m(G, k/2) \text{ if } k \text{ is even.}$$

Furthermore

$$Ch(G, x) = \sum_{k=even} (-1)^{k/2} m(G, k/2) x^{n-k}.$$

The right hand side of the above relation is obviously the matching polynomial of G.

In order to complete the proof of 4.5.2.1, it remains to be shown that the above relation does not hold if G contains cycles.

If one applies Sachs' theorem 4.4.3.2 to the case $k = g$, where g is the size of the smallest cycle in G, a similar argument to that already used here and in 4.4.3.4 gives the results:

$$c(g) = -2n_g \qquad \text{if } g \text{ is odd}$$

$$c(g) = (-1)^{g/2} m(G, g/2) - 2n_g \quad \text{if } g \text{ is even,}$$

where n_g is the number of cycles of size g contained in the graph G as subgraphs. Consequently, the matching and the characteristic polynomial will not coincide whenever G is cyclic (i.e., when $n_g \neq 0$).

This completes the proof of both Theorems 4.5.2.1 and 4.4.1.9.

If G is a cyclic graph, the relations between $\text{Ma}(G)$ and $\text{Ch}(G)$ are somewhat more complicated.

4.5.2.3. THEOREM. (Hosoya [44]) *Let C be a regular graph of degree two with $p(C)$ components. Then*

$$\text{Ch}(G) = \text{Ma}(G) + \sum_C (-2)^{p(C)} \text{Ma}(G - C)$$

with the summation extending over all regular graphs of degree two contained in G.

4.5.2.4. COROLLARY. *Let the graph G contain the cycles $Z_a, Z_b, \ldots, Z_t$. Then*

$$\text{Ch}(G) = \text{Ma}(G) - 2\sum_a \text{Ma}(G - Z_a) + 4\sum_{a,b} \text{Ma}(G - Z_a - Z_b) -$$

$$- 8\sum_{a,b,c} \text{Ma}(G - Z_a - Z_b - Z_c) + \ldots .$$

The first summation of the r.h.s. of the above equation extends over all cycles of G; the second, third etc. summations extend over all pairs, triplets etc. of mutually independent cycles of G.

4.5.2.5. THEOREM. (Gutman [83])

$$\text{Ma}(G) = \text{Ch}(G) + \sum_C (+2)^{p(C)} \text{Ch}(G - C)$$

where symbols have the same meaning as in Theorem 4.5.2.3.

4.5.2.6. COROLLARY.

$$Ma(G) = Ch(G) + 2\sum_{a} Ch(G - Z_a) + 4\sum_{a,b} Ch(G - Z_a - Z_b) +$$

$$+ 8\sum_{a,b,c} Ch(G - Z_a - Z_b - Z_c) + \dots .$$

As an illustration, we now compute $Ch(G_{61})$ using the result 4.5.2.4. The graph G_{61} contains six cycles, Z_1 to Z_2, of which only the cycles Z_1 and Z_3 are independent.

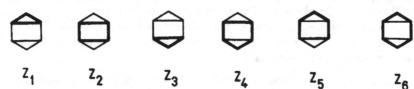

$$Z_1 \qquad Z_2 \qquad Z_3 \qquad Z_4 \qquad Z_5 \qquad Z_6$$

Therefore

$$Ch(G_{61}) = Ma(G_{61}) - 2\sum_{i=1}^{6} Ma(G_{61} - Z_i) + 4Ma(G_{61} - Z_1 - Z_3).$$

$Ch(G_{61})$ can be immediately obtained from the previously determined $Ma(G_{61})$ and the simple relations:

$$Ma(G_{61} - Z_1) \qquad = Ma(G_{61} - Z_3) = x^3 - 3x$$

$$Ma(G_{61} - Z_2) \qquad = x^2$$

$$Ma(G_{61} - Z_4) \qquad = Ma(G_{61} - Z_5) = x$$

$$Ma(G_{61} - Z_6) \qquad = 1$$

$$Ma(G_{61} - Z_1 - Z_3) = 1.$$

After appropriate algebraic manipulations we get:

$$Ch(G_{61}) = x^6 - 8x^4 - 4x^3 + 12x^2 + 2x.$$

A few years ago Godsil [84] discovered another relation between Ma and Ch. This seems to be a crucial result in the theory of the matching polynomial.

4.5.2.7. THEOREM. (Godsil) *For every connected graph* G *there exists a tree* $T = T(G)$, *such that* Ma(G) *is a divisor of* Ch(T).

The tree $T(G)$ may be called the Godsil tree of the graph G. The Godsil tree is uniquely determined by the graph G *and* its (arbitrary) vertex v.

The construction of a Godsil tree is not difficult. It is based on a consideration of all the paths in G connecting the vertex v with other vertices. We shall not go into the details here of the proof of Godsil's theorem or of the structure of the Godsil tree. The interested reader is referred to [84,85,157]. We mention also that Godsil trees were investigated by Randić *et al* [86] in connection with quite a different problem.

The Godsil tree of the cycle C_n is the path P_{2n-1} (irrespective of the starting vertex v). It can be proved that:

$$Ch(P_{2n-1}) = Ma(C_n) \cdot Ch(P_{n-1}).$$

We see that $Ch(P_{2n-1})$ has two divisors: $Ma(C_n)$ and $Ch(P_{n-1})$.

The main consequence of Theorem 4.5.2.7 is the following rather important result.

4.5.2.8. THEOREM. *All the zeros of the matching polynomial (of a simple graph) are real numbers.*

4.5.2.9. PROOF OF 4.5.2.8. Let T be the Godsil tree of a graph G. According to 4.4.2.5 all the zeros of Ch(T) are real numbers. Since Ma(G) is a divisor of Ch(T), all the zeros of Ma(G) are zeros of Ch(T).

The result 4.5.2.8 was proved long before Godsil's theorem 4.5.2.7 was put forward. As already mentioned, 4.5.2.8 was first deduced in 1970 by Heilmann and Lieb [64], and independently by Kunz [65]. In the paper [67] Heilmann and Lieb even offer three different proofs of 4.5.2.8.

Without knowledge of references [64] to [67], Godsil and Gutman proved the realness of the zeros of Ma(x) in 1978 [87–89]. Note, however, that all the previous proofs were much more complicated than that in 4.5.2.9. For a full bibliography on the history of Theorem 4.5.2.8 see reference [76]. Theorem 4.5.2.8 has been extended to graphs having arbitrarily weighted edges and self-loops [10].

Another approach towards the proof of the real nature of the zeros of Ma(x) was that of Herndon *et al* [90,91] and of Aihara [92]. Herndon showed that in certain cases a graph G^R (not necessarily simple) can be constructed, such that $\mathrm{Ch}(G^R) = \mathrm{Ma}(G)$. Aihara demonstrated that in certain cases a Hermitian matrix H (with complex matrix elements) can be constructed, such that $\mathrm{Ch}(\mathrm{H}) = \mathrm{Ma}(G)$. Although the method of Aihara was later further elaborated [93–95], it has proved [96] that neither G^R nor H exist in the general case (even in the case of some bicyclic graphs).

The main merit of Theorem 4.5.2.7 is that it reduces the entire theory of the matching polynomial to the much more developed graph spectral theory. It is thus now no longer surprising that there are numerous analogies between $\mathrm{Ma}(G)$ and $\mathrm{Ch}(G)$. We list two more results of this kind [97–98], and refer the interested reader to reference [99].

4.5.2.10. THEOREM. (analogous to 4.5.5.2) *If $v_1, v_2, \ldots, v_n$ are the vertices of* G, *then*

$$\frac{d\mathrm{Ma}(G, x)}{dx} = \sum_{i=1}^{n} \mathrm{Ma}(G - v_i, \ x).$$

4.5.2.11. THEOREM. (analogous to 4.4.4.7) *Let P_{rs} be a path in the graph* G *connecting the vertices v_r and v_s. Then*

$$\mathrm{Ma}(G - v_r) \cdot \mathrm{Ma}(G - v_s) - \mathrm{Ma}(G) \cdot \mathrm{Ma}(G - v_r - v_s) = \sum [\mathrm{Ma}(G - P_{rs})]^2$$

with the summation on the r.h.s. ranging over all paths in G *connecting v_r with v_s.*

4.5.3 The Matching Polynomial and Sachs' Theorem

One way in which the matching polynomial was discovered [68–70,72] was to modify Sachs' theorem (4.4.3.2 and 4.4.3.3). Thus, instead of Definition 4.5.1.3 we can introduce the matching polynomial via the following statement resembling Sachs' theorem:

4.5.3.1. DEFINITION. (analogous to 4.4.3.3)

$$\mathrm{Ma}(G, x) = x^n + \sum_{s} (-1)^{p(s)} 2^{c(s)} x^{n - n(s)}$$

where the summation extends over all Sachs' graphs s contained as subgraphs in the graph G, *and possessing no cyclic components. In other* words, contrary to Theorem 4.4.3.3, we restrict our summation to Sachs graphs s for which $c(s) = 0$.

4.5.3.2. THEOREM. *Definitions 4.5.1.3 and 4.5.3.1 are equivalent.*

4.5.3.3. PROOF OF 4.5.3.2. Definition 4.5.3.1 can be written as

$$Ma(G, x) = x^n + \sum_s (-1)^{p(s)} x^{n-n(s)}$$

because only Sachs' graphs with the property $c(s) = 0$ are taken into account. Such Sachs' graphs necessarily possess an even number of vertices and all their components are K_2 graphs. Therefore $p(s) = n(s)/2$. Every such Sachs' graph is in a one-to-one correspondence with a $p(s)$-matching in G. The number of acyclic Sachs' graphs with p components is thus equal to the number of p-matchings of G. Theorem 4.5.3.2 follows.

From Definition 4.5.3.1 it is obvious why the name "acyclic polynomial" has been chosen [68–71] for Ma. This definition (or more precisely, Theorem 4.5.3.2) is indispensable in the theory of topological resonance energy (TRE). We can reformulate Theorem 4.5.3.2 in the following manner.

4.5.3.4. COROLLARY. *The matching polynomial is obtained from the characteristic polynomial by neglecting the contributions coming from cyclic Sachs' graphs.*

Another consequence of Theorem 4.5.3.2 is of some importance in theoretical chemistry and presents the basis of the TRE concept [100,101].

4.5.3.5. COROLLARY. (Gutman) *Let a physical molecular quantity Q be computed from the characteristic polynomial of the molecular graph G according to the mapping f:*

$$Ch(G) \xrightarrow{f} Q.$$

Then the quantity Q^R, defined by means of

$$Ma(G) \xrightarrow{f} Q^R$$

can be interpreted as an analogy of the quantity Q, in which the effect of cyclic conjugation has been neglected. Accordingly, $Q - Q^R$ should be interpreted as the effect of cyclic conjugation on the quantity Q.

In particular, if Q is chosen to the the total π-electron energy (E_π), E_π^R can be interpreted as the total π-electron energy of the reference structure, and the difference $E_\pi - E_\pi^R$ is a measure of the effect of cyclic conjugation on the stability of the conjugated molecule considered. This difference was named the "topological resonance energy."

4.5.4 Chemical Applications of Matching Polynomials

The theory of topological resonance energy (TRE) represents an important area for chemical applications of matching polynomials. As this theory will be discussed in more detail elsewhere in this book, we offer here only few brief remarks. As already explained in 4.5.3.5, TRE is the measure of the effect of cyclic conjugation [100,101] on total π-electron energy. TRE therefore also measures the effect of cyclic conjugation on the thermodynamic stability of a conjugated molecule (within the Hückel molecular orbital approximation). This interpretation of TRE is beyond dispute.

On the other hand, TRE was also proposed as a quantitative topological measure of aromaticity [68,70,72]. Since the aromaticity of a conjugated molecule depends on many factors – and not only on cyclic conjugation – it is hardly surprising that this second application of TRE led to many controversies. In the recent chemical literature TRE has been criticized by some authors [102–110] and defended by others [111–113].

As mentioned, Hosoya used the polynomial $M(x)$ (see Definition 4.5.1.4) in his topological theory of saturated hydrocarbons [43]. The success of Hosoya's theory is based on the fact [114] that the absolute entropy of an acyclic saturated hydrocarbon is linearly correlated with the logarithm of the topological index Z_G. Hosoya's index $Z_G = M(G,1)$ is, of course, just the total number of matchings in the graph G. Some mathematical properties of Z_G were derived in references [115–117] and [163–165]. The topological index [118,119] and related quantities [120–122] were applied to the theory of conjugated hydrocarbons. Hosoya also proposed Z_G as a sorting device for the coding of chemical structures [123].

4.6 More Graphic Polynomials

In this section we shall briefly describe some additional graphic polynomials which are of relevance in chemical applications. Readers interested in further details should consult the literature cited.

4.6.1 The Mülheim polynomial

Numerous examples given in the previous two sections exhibit the profound and far-reaching analogy existing between the characteristic and matching polynomials of a graph.

The question has been raised whether a continuous mapping between $Ma(x)$ and $Ch(x)$ can be constructed [102,124,125].

The idea of partial neglect of cyclic Sachs' graphs in Sachs' formula (4.4.3.2) led to the discovery of a polynomial which generalizes both $Ch(x)$ and $Ma(x)$. Since the idea was born in the Institut für Strahlenchemie in Mülheim (F.R.G.) during a discussion between I. Gutman and O.E. Polansky, the name "Mülheim polynomial" seems appropriate.

4.6.1.1. DEFINITION. *Let the graph G contain the cycles $Z_1, Z_2, \ldots, Z_t$. Let weight (i.e., a real number) T_i be associated with each cycle Z_i, with $i - 1, \ldots, t$, and let $T = (T_1, T_2, \ldots, T_t)$. Let s be a Sachs' graph of G containing the cycles $Z_a, Z_b, \ldots, Z_p$. Then we define $T(s) = T_a \cdot T_b \ldots T_p$. If s is acyclic, we set $T(s) = 1$.*

The Mülheim polynomial $Mu(G)$ of the graph G is defined now in analogy to that in Sachs' theorem 4.4.3.2. The weight T_i measures that extent to which the cycle Z_i will be taken into account in the Sachs-type summation.

4.6.1.2. DEFINITION. *If the Mülheim polynomial is written in the form:*

$$Mu(G, T, x) = x^n + \sum_{k=1}^{n} mu(G, T, k)x^{n-k}$$

we have

$$Mu(G, T, k) = \sum_{s}(-1)^{p(s)}2^{c(s)}; \ T(s),$$

where the summation extends over all Sachs' graphs on k vertices contained in the graph G.

Its mathematical properties [101,126,127], a more general definition [128], and chemical applications [101,102,127,129–131,158,159] will not be discussed here. We shall mention, however, two important consequences of 4.6.1.2.

4.6.1.3. COROLLARY. *If $T = (0, 0, \ldots, 0)$, then $Mu(G, T, x) = Ma(G, x)$ and if $T = (1, 1, \ldots, 1)$, then $Mu(G, T, x) = Ch(G, x)$. Hence, by varying the components of the vector T one can continuously map Ma into Ch and vice versa.*

4.6.1.4. COROLLARY.

$$\frac{\partial}{\partial T_a} \, \mathrm{Mu}(G, \mathrm{T}, x) = \mathrm{Mu}(G - Z_a, \mathrm{T}, x).$$

4.6.2 Effect of a Single Cycle

If one neglects only the contributions of a single cycle Z in Sachs' formula 4.4.3.2, a polynomial $\mathrm{Si}(G, Z, x)$ will be obtained [132].

4.6.2.1. THEOREM.

$$\mathrm{Si}(G, Z, x) = \mathrm{Ch}(G, x) + 2\mathrm{Ch}(G - Z, \ x).$$

4.6.2.2. THEOREM. $\mathrm{Si}(G, Z, x) = \mathrm{Mu}(G, \mathrm{T}, x)$ *for the choice* $T_i = 0$ *for* $Z_i = Z$ *and* $T_i = 1$ *otherwise.*

The polynomial $\mathrm{Si}(G, Z, x)$ had been used for calculating the effect of a cycle on the thermodynamic stability [132–134] and π-electron charge distribution [135] of a conjugated molecule.

Herndon [136] pointed out that the fact that the zeros of $\mathrm{Si}(G, Z, x)$ are not necessarily real numbers may cause certain difficulties in the chemical applications of this polynomial. These problems have recently been overcome [156].

4.6.3 The Independence Polynomial

In full analogy to Definition 4.5.1.3, we now introduce the independence polynomial as follows.

4.6.3.1. DEFINITION.

$$\mathrm{In}(G, x) = \sum_{k=0}^{n} (-1)^k n(G, k) x^{n-2k},$$

where $n(G, k)$ is the number of selections of k independent vertices in the graph G. In addition, $n(G, 0) = 1$ and $n(G, 1) = n =$ the number of vertices of G.

The independence polynomial generalizes a great number of graphic polynomials. For example, one can verify the following statements.

4.6.3.2. THEOREM. *Every matching polynomial is an independence polynomial.*

4.6.3.3. THEOREM. *Every rook polynomial is an independence polynomial.*

4.6.3.4. THEOREM. (Gutman) *Every sextet polynomial is an independence polynomial.*

Readers interested in rook polynomials should consult references [75] and [76]. The sextet polynomial will be introduced later on.

The properties of the independence polynomial are to some extent analogous to the properties of the matching polynomial. However, in the general case the zeros of $In(G)$ are not real numbers. The mathematical theory of $In(G)$ was developed by Gutman and Harary [137]. Some applications of $In(G)$ to chemical problems have been attempted [138,139].

4.6.4 Polynomials Associated with Benzenoid Systems

Benzenoid systems have especially interesting topological properties. Many of these properties are reflected in the corresponding graphic polynomials [109,140]. As an example, we mention here the famous result of Dewar and Longuet-Higgins [141].

4.6.4.1. THEOREM. (Dewar and Longuet-Higgins) *If G is the molecular graph of a benzenoid system, then*

$$Ch(G,0) = (-1)^{n/2} K^2$$

where K is the number of Kekulé structures in the corresponding molecule.

A recent result [109] is the relation between the coefficients of Ma and Ch.

4.6.4.2. THEOREM. (Gutman) *If G is the molecular graph of a benzenoid system, then for all k,*

$$M(G,k) \le |c(G,2k)| \le m(G,k)^2.$$

The definition of a Clar resonant sextet formula (of a benzenoid hydrocarbon) can be found elsewhere [63,142]. Let G be the molecular graph of a benzenoid hydrocarbon and let $s(G,k)$ be the number of Clar formulas of G with k sextets.

4.6.4.3.   DEFINITION. (Hosoya and Yamaguchi [142])   *The sextet polynomial of G is defined as:*

$$Sex(G) = Sex(G,x) = \sum_{k=0}^{S} s(G,k)x^{k},$$

where $s(G,O) = 1$ and S is the maximal number of resonant sextets in G.

The theory of the sextet polynomial is currently rapidly developing [139,142-151,173-178]. One of the most exciting discoveries in this field was made in [142] and proved in [146].

4.6.4.4.   THEOREM. *If G is a catacondensed system, then*

$$Sex(G,1) = K,$$

where K is the number of Kekulé structures of G.

The above relation holds for some, but not for all pericondensed benzenoids. Necessary and sufficient conditions which a pericondensed benzenoid system must obey in order that $Sex(G,1) = K$ have been recently established by Zhang and Chen [179].

4.6.4.5.   THEOREM. (Gutman [145]) *If G is an unbranched catacondensed benzenoid system, there exists a tree $T = T(G)$, such that*

$$Sex(G,x) = M(T(G),x),$$

where M denotes the matching polynomial defined in 4.5.1.4. $T(G)$ is sometimes called "Gutman tree" [177,178].

The finding that for every benzenoid graph G another graph $C(G)$ can be constructed [139], such that $Sex(G) = \ln(C(G))$, had already been mentioned (Theorem 4.6.3.4). The graph $C(G)$ is of some relevance in Clar's aromatic sextet theory. The name "Clar graph" has been proposed for $C(G)$.

More details on the topological properties of benzenoid systems (including their graphic polynomials) can be found in the reviews [63,178].

4.6.5   Miscellaneous

A few other graphic polynomials have been considered in the chemical literature. We would like just to mention in concluding the permanental [152], the king [172], the Clar [176], the cyclic [38], the homomorphism

[198] polynomial and the polynomials proposed by Wheland [153,154,199], Knop and Trinajstić [38,39] and Gutman [155,200]. Their chemical relevance remains to be justified in the future.

4.7 References

The following list of references is far from being complete. A neophyte, intending to become more familiar with graphic polynomials should first consult the textbooks and reviews marked by an asterisk, and thereafter the original scientific papers marked by two asterisks. Readers interested in chemical applications of graphic polynomials should also consult the other chapters of the present book.

1. *D.M. Cvetković, M. Doob and H. Sachs, *Spectra of Graphs – Theory and Application*, Academic Press, New York 1980.
2. A. Graovac, O.E. Polansky, N. Trinajstić and N.N. Tyutyulkov, *Z. Naturforsch.* 30a (1975), 1696.
3. **J. Aihara, J. Amer. Chem. Soc. 98 (1976), 2750.
4. M.J. Rigby, R.B. Mallion and A.C. Day, Chem. Phys. Lett. 51 (1977), 178.
5. D.A. Bochvar, I.V. Stankevich and A.L. Chistyakov, Zh. Fiz. Khim. 35 (1961), 1337.
6. R.B. Mallion, N. Trinajstić and A.J. Schwenk, Z. Naturforsch. 29a (1974), 1481.
7. R.B. Mallion, A.J. Schwenk and N. Trinajstić, Croat. Chem. Acta 46 (1974), 171.
8. M.J. Rigby and R.B. Mallion, J. Comb. Theory B27 (1979), 122.
9. W.C. Herndon and M.L. Ellzey, J. Chem. Inf. Comput. Sci. 19 (1979), 260.
10. I. Gutman, Croat. Chem. Acta 54 (1981), 75.
11. D.A. Bochvar, I.V. Stankevich and A.L. Chistyakov, Zh. Fiz. Khim. 39 (1965), 1365.
12. D.A. Bochvar, I.V. Stankevich, Zh. Fiz. Khim. 39 (1965), 2028.
13. **I.Gutman, Theoret. Chim. Acta 50 (1979), 287.
14. *R.B. Mallion, in: R.J. Wilson (Ed.), *Applications of Combinatorics*, Shiva Publ., Nantwich 1982, pp. 87–114.
15. L.N. Lichtenbaum, in: *Trudy 3. Vses. Matem. S'ezda, Vol. 1*, 1956, pp. 135–136 (cited according to Ref. 1).

16. L. Collatz and U. Sinogowitz, Ahb. Math. Sem. Univ. Hamburg 21 (1957), 63.
17. *S.S. D'Amato, B.M. Gimarc and N. Trinajstić, Croat. Chem. Acta 54 (1981), 1.
18. A.J. Schwenk, in: F. Harary (Ed.), *New Directions in the Theory of Graphs*, Academic Press, New York 1973, pp. 275–307.
19. E. Heilbronner and T.B. Jones, J. Amer. Chem. Soc. 100 (1978), 6506.
20. A.T. Balaban and F. Harary, J. Chem. Docu. 11 (1971), 258.
21. I.M. Mladenov, M.D. Kotarov and J.G. Vassileva-Popova, *Int. J. Quantum Chem.* 18 (1980), 339.
22. C.A. Coulson and G.S. Rushbrooke, Proc. Cambridge Phil. Soc. 36 (1940), 193.
23. D. Cvetković, Matematička Biblioteka (Beograd) 41 (1969), 193.
24. H. Sachs, Publ. Math. (Debrecen) 11 (1963), 119.
25. F. Harary, SIAM Rev. 4 (1962), 202.
26. N. Milić, IEEE Trans. Circuit Theory CT-11 (1964), 423.
27. L. Spialter, J. Chem. Docu. 4 (1964), 269.
28. A. Graovac, I. Gutman, N. Trinajstić and T. Živković, Theoret. Chim. Acta 26 (1972), 67.
29. D. Cvetković, I. Gutman and N. Trinajstić, Croat. Chem. Acta 44 (1972), 365.
30. D. Cvetković, I. Gutman and N. Trinajstić, Chem. Phys. Letters 16 (1972), 614.
31. *I.Gutman and N.Trinajstić, Topics Curr. Chem. 42 (1973), 49.
32. I. Gutman and N. Trinajstić, Croat. Chem. Acta 45 (1973), 423.
33. *I. Gutman and N. Trinajstić, Croat. Chem. Acta 47 (1975), 507.
34. N. Trinajstić, Croat. Chem. Acta 49 (1977), 593.
35. *A. Graovac, I. Gutman and N. Trinajstić, *Topological Approach to the Chemistry of Conjugated Molecules*, Springer-Verlag, Berlin 1977.
36. I.S. Dmitriev, *Molekuly bez khimicheskih svyazei*, Khimia, Leningrad 1980. English translation: I.S. Dmitriev, *Molecules Without Chemical Bonds*, Mir, Moscow 1981.
37. F.H. Clarke, Discrete Math. 3 (1972), 305.
38. J.V. Knop and N. Trinajstić, Int. J. Quantum Chem. Symp. 14 (1980), 503.
39. I. Gutman, Croat. Chem. Acta 55 (1982), 309.
40. M. Randić, J. Comput. Chem. 3 (1982), 421.
41. E. Heilbronner, Helv. Chim. Acta 36 (1953), 170.

42. A.J. Schwenk, in: R. Bary and F. Harary (Eds.), *Graphs and Combinatorics*, Springer-Verlag, Berlin 1974, pp. 153–172.
43. H. Hosoya, Bull. Chem. Soc. Japan 44 (1971), 2332.
44. H. Hosoya, Theoret. Chim. Acta 25 (1972), 215.
45. M.V. Kaulgud and V.H. Chitgopkar, J. Chem. Soc. Faraday II 73 (1977), 1385.
46. M.V. Kaulgud and V.H. Chitgopkar, J. Chem. Soc. Faraday II 74 (1978), 951.
47. I. Gutman, J. Chem. Soc. Faraday II 76 (1980), 1161.
48. I. Gutman, Publ. Inst. Math. (Beograd) 21 (1977), 75.
49. Tang Au-chin and Kiang Yuan-sun, Sci. Sinica 19 (1976), 207.
50. Zhang Fuji, Sci. Sinica 22 (1979), 1160.
51. Yan Guo-sen, Int. J. Quantum Chem. Symp. 14 (1980), 549.
52. Yuan-sun Kiang, Int. J. Quantum Chem. Symp. 15 (1981), 293.
53. Jiang Yuansheng, Sci. Sinica 25 (1982), 681.
54. K. Balasubramanian and M. Randić, Theoret. Chim. Acta 61 (1982), 307.
55. K. Balasubramanian, Int. J. Quantum Chem. 21 (1982), 581.
56. C.A. Coulson and H.C. Longuet-Higgins, Proc. Roy. Soc. (London) A192 (1947), 16.
57. I. Gutman, Z. Naturforsch. 36a (1981), 1112.
58. L. Spialter, J. Amer. Chem. Soc. 85 (1963), 2012.
59. L. Lovász and J. Pelikán, Period. Math. Hung. 3 (1973), 175.
60. I. Gutman, B. Ruščić, N. Trinajstić and C.F. Wilcox, J. Chem. Phys. 62 (1975), 3399.
61. D.M. Cvetković and I. Gutman, Croat. Chem. Acta 49 (1979), 115.
62. W.C. Herndon, J. Chem. Educ. 51 (1974), 10.
63. *I. Gutman, Bull. Soc. Chim. Beograd 47 (1982), 453.
64. O.J. Heilmann and E.H. Lieb, Phys. Rev. Lett. 24 (1970), 1412.
65. H. Kunz, Phys. Letters A32 (1970), 311.
66. C. Gruber and H. Kunz, Commun. Math. Phys. 22 (1971), 133.
67. **O.J. Heilmann and E.H. Lieb, Commun. Math. Phys. 25 (1972), 190.
68. I. Gutman and N. Trinajstić, Acta Chim. Acad. Sci. Hung. 91 (1976), 203.
69. I. Gutman, M. Milun and N. Trinajstić, Croat. Chem. Acta 48 (1976), 87.
70. I. Gutman, M. Milun and N. Trinajstić, J. Amer. Chem. Soc. 99 (1977), 1692.
71. I. Gutman, Publ. Inst. Math. (Beograd) 22 (1977), 63.

72. J. Aihara, J. Amer. Chem. Soc. 98 (1976), 2750.
73. E.J. Farrell, J. Comb. Theory B27 (1979), 75.
74. E.J. Farrell, Ars Combinatoria 9 (1980), 221.
75. J. Riordan, *An Introduction to Combinatorial Analysis*, Wiley, New York 1958, Chapters 7 and 8.
76. C.D. Godsil and I. Gutman, Croat. Chem. Acta 54 (1981), 53.
77. *I.Gutman, Mth. Chem. 6 (1979), 75.
78. **C.D. Godsil and I. Gutman, J. Graph Theory 5 (1981), 137.
79. I. Gutman and D.N. Cvetković, Coll. Sci. Papers Fac. Sci. Kragujevac 1 (1980), 101.
80. H. Hosoya, Nat. Sci. Rept. Ochanomizu Univ. 32 (1981), 127.
81. C.D. Godsil, Combinatorica 1 (1981), 257.
82. T. Zaslavsky, Europ. J. Comb. 2 (1982), 91.
83. I. Gutman, Ph.D. Thesis (Mathematica), University of Belgrade, 1980.
84. C.D. Godsil, J. Graph Theory 5 (1981), 285.
85. **C.D. Godsil and I. Gutman, Acta Chim. Acad. Sci. Hung. 110 (1982), 415.
86. M. Randić, G.N. Brissey, R.B. Spencer and C.L. Wilkins, Comput. Chem. 3 (1979), 5.
87. C.D. Godsil and I. Gutman, Math. Research Report Univ. Melbourne 35 (1978), 1.
88. C.D. Godsil and I. Gutman, Z. Naturforsch. 34a (1979), 776.
89. C.D. Godsil and I. Gutman, in: L Lovász and V.T. Sós (Eds.), *Algebraic Methods in Graph Theory*, North-Holland, Amsterdam 1981, pp. 241–249.
90. W.C. Herndon and C. Párkányi, Tetrahedron 34 (1978), 3419.
91. W.C. Herndon and M.L. Ellzey, J. Chem. Inf. Comput. Sci. 19 (1979), 260.
92. J. Aihara, Bull. Chem. Soc. Japan 52 (1979), 1529.
93. L.J. Schaad, B.A. Hess, J.B. Nation and N. Trinajstić (with an appendix by I. Gutman), Croat. Chem. Acta 52 (1979), 233.
94. A. Graovac, Chem. Phys. Lett. 82 (1981), 248.
95. A. Graovac, D. Kasum and N. Trinajstić, Croat. Chem. Acta 54 (1981), 91.
96. I. Gutman, A. Graovac and B. Mohar, Math. Chem. 13 (1982), 129.
97. I. Gutman and H. Hosoya, Theoret. Chim. Acta 48 (1978), 279.
98. C.D. Godsil, Research Report 82-08, Simon Fraser Univ., Burnaby 1982.
99. I. Gutman, Publ. Inst. Math. (Beograd) 31 (1982), 27.

100. I. Gutman, Croat. Chem. Acta 53 (1980), 581.
101. **I. Gutman and O.E. Polansky, Theoret. Chim. Acta 60 (1981), 203.
102. I. Gutman, Chem. Phys. Lett. 66 (1979), 595.
103. I. Gutman, Theoret. Chim. Acta 56 (1980), 89.
104. I. Gutman and B. Mohar, Chem. Phys. Lett. 69 (1980), 375.
105. I. Gutman and B. Mohar, Chem. Phys. Lett. 77 (1981), 567.
106. E. Heilbronner, Chem. Phys. Lett. 85 (1982), 377.
107. W.C. Herndon, J. Org. Chem. 46 (1981), 2119.
108. I. Gutman and B. Mohar, Croat. Chem. Acta 55 (1982), 375.
109. I. Gutman, J. Chem. Soc. Faraday II 79 (1983), 337.
110. I. Gutman and A.V. Teodorović, Bull. Soc. Chim. Beograd 47 (1982), 579.
111. J. Aihara, Chem. Phys. Lett. 73 (1980), 404.
112. P. Ilić and N. Trinajstić, J. Org. Chem. 45 (1980), 1738.
113. A. Sabljić and N. Trinajstić, J. Org. Chem. 46 (1981), 3457.
114. H. Narumi and H. Hosoya, Bull. Chem. Soc. Japan 53 (1980), 1228.
115. H. Hosoya, Fibonacci Quart. 11 (1973), 255.
116. I. Gutman, Theoret. Chim. Acta 45 (1977), 79.
117. I. Gutman, Croat. Chem. Acta 54 (1981), 81.
118. H. Hosoya and M. Murakami, Bull. Chem. Soc. Japan 48 (1975), 3512.
119. H. Hosoya and K. Hosoi, J. Chem. Phys. 64 (1976), 1065.
120. H. Hosoya, K. Hosoi and I. Gutman, Theoret. Chim. Acta 38 (1975), 37.
121. I. Gutman, T. Yamaguchi and H. Hosoya, Bull. Chem. Soc. Japan 49 (1976), 1811.
122. J. Aihara, J. Org. Chem. 41 (1976), 2488.
123. H. Hosoya, J. Chem. Docum. 12 (1972), 181.
124. J. Aihara, Bull. Chem. Soc. Japan 50 (1977), 2010 (Appendix).
125. J. Aihara, Bull. Chem. Soc. Japan 51 (1978), 1788.
126. O.E. Polansky and A. Graovac, Math. Chem. 13 (1982), 151.
127. O.E. Polansky and M. Zander, J. Mol. Struct. 84 (1982), 361.
128. E.J. Farrell, J. Comb. Theory 26B (1979), 111.
129. I. Gutman, Z. Naturforsch. 35a (1980), 458.
130. I. Gutman, Bull. Soc. Chim. Beograd 44 (1979), 627.
131. I. Gutman and I. Juranić, Bull. Soc. Chim. Beograd 47 (1982), 183.
132. S. Bosanac and I. Gutman, Z. Naturforsch. 32a (1977), 10.
133. I. Gutman and S. Bosanac, Tetrahedron 33 (1977), 1809.
134. I. Gutman, J. Chem. Soc. Faraday II 75 (1979), 799.

135. I. Gutman, Coll. Sci. Papers Fac. Sci. Kragujevac 2 (1981), 157.
136. W.C. Herndon, J. Amer. Chem. Soc. 104 (1982), 3541.
137. I. Gutman and F. Harary, Utilitas Math. 24 (1983), 97.
138. R.E. Merrifield and H.E. Simmons, Theoret. Chim. Acta 55 (1980), 55.
139. I. Gutman, Z. Naturforsch. 37a (1982), 69.
140. I. Gutman, Croat. Chem. Acta 46 (1974), 209.
141. M.J.S. Dewar and H.C. Longuet-Higgins, Proc. Roy. Soc. (London) A214 (1952), 482.
142. **H. Hosoya and T. Yamaguchi, Tetrahedron Lett. (1975), 4659.
143. J. Aihara, Bull. Chem. Soc. Japan 49 (1976), 1429.
144. J. Aihara, Bull. Chem. Soc. Japan 50 (1977), 2010.
145. I. Gutman, Theoret. Chim. Acta 45 (1977), 309.
146. I. Gutman, H. Hosoya, T. Yamaguchi, A. Motoyama and N. Kuboi, Bull. Soc. Chim. Beograd 42 (1977), 503.
147. I. Gutman, Math. Chem. 11 (1981), 127.
148. N. Ohkami, A. Motoyama, T. Yamaguchi, H. Hosoya and I. Gutman, Tetrahedron 37 (1981), 1113.
149. S. El-Basil, Chem. Phys. Lett. 89 (1982), 145.
150. S. El-Basil and I. Gutman, Chem. Phys. Lett. 94 (1983), 188.
151. S. El-Basil, Croat. Chem. Acta 57 (1984), 1.
152. D. Kasum, N. Trinajstić and I. Gutman, Croat. Chem. Acta 54 (1981), 321.
153. G.W. Wheland, J. Chem. Phys. 3 (1935), 356.
154. N. Ohkami and H. Hosoya, Bull. Chem. Soc. Japan 52 (1979), 1642.
155. I. Gutman, Bull. Soc. Chim. Beograd 46 (1981), 17.
156. I. Gutman and W.C. Herndon, Chem. Phys. Lett. 105 (1984), 281.
157. *D. Cvetković, M. Doob, I. Gutman and A. Torgašev, *Recent Results in the Theory of Graph Spectra*, North Holland, Amsterdam 1988.
158. I. Gutman, Theoret. Chim. Acta 66 (1984), 43.
159. I. Gutman, Chem. Phys. Lett. 117 (1985), 614.
160. *S.J. Cyvin and I. Gutman, Comp. & Math. with Appl. 12B (1986), 859.
161. *S.J. Cyvin and I. Gutman, *Kekulé Structures in Benzenoid Hydrocarbons*, Springer-Verlag, Berlin 1988.
162. *I. Gutman and O.E. Polansky, *Mathematical Concepts in Organic Chemistry*, Springer-Verlag, Berlin 1986.
163. I. Gutman and J. Cioslowski, Z. Naturforsch. 42a (1987), 438.
164. I. Gutman and Z. Marković, Bull. Chem. Soc. Japan 60 (1987), 2611.

165. I. Gutman, Z. Marković and S. Marković, Chem. Phys. Lett. 134 (1987), 139.
166. A. Graovac and O.E. Polansky, Math. Chem. 21 (1986), 33.
167. O.E. Polansky and A. Graovac, Math. Chem. 21 (1986), 47.
168. A. Graovac and O.E. Polansky, Math. Chem. 21 (1986), 81.
169. O.E. Polansky and A. Graovac, Math. Chem. 21 (1986), 93.
170. I. Gutman, Math. Chem. 19 (1986), 127.
171. I. Gutman and O.E. Polansky, Math. Chem. 19 (1986), 139.
172. K. Balasubramanian and R. Ramaraj, J. Comput. Chem. 6 (1985), 447.
173. S. El-Basil, Theoret. Chim. Acta 65 (1984), 191.
174. S. El-Basil, Theoret. Chim. Acta 65 (1984), 199.
175. I. Gutman and S. El-Basil, Z. Naturforsch. 40a (1985), 923.
176. S. El-Basil, Theoret. Chim. Acta 70 (1986), 53.
177. S. El-Basil, J. Chem. Soc. Faraday II 82 (1986), 299.
178. *S. El-Basil, J. Math. Chem. 1 (1987), 153.
179. Zhang Fuji and Chen Rong-si, Math. Chem. 19 (1986), 179.
180. *K. Balasubramanian, Chem. Rev. 85 (1985), 599.
181. J.R. Dias, Canad. J. Chem. 65 (1987), 734.
182. M. Randić and B. Baker, J. Math. Chem. 2 (1988), 241.
183. J.R. Dias, J. Chem. Educ. 64 (1987), 213.
184. I. Gutman, J. Math. Chem. 1 (1987), 123.
185. N. Trinajstić, J. Math. Chem. 2 (1988), 197.
186. J.R. Dias, Theor. Chim. Acta 68 (1985), 107.
187. J.R. Dias, Nouv. J. Chim. 9 (1985), 125.
188. J.R. Dias, J. Mol. Struct. (Theochem) 149 (1987), 213.
189. J.R. Dias, *Handbook of Polycyclic Hydrocarbons. Part A. Benzenoid Hydrocarbons*, Elsevier, Amsterdam 1987.
190. K. Balasubramanian, J. Comput. Chem. 6 (1985), 656.
191. M. Randić, J. Math. Chem. 1 (1987), 145.
192. K. Balasubramanian, J. Comput. Chem. 9 (1988), 204.
193. K. Balasubramanian, J. Comput. Chem. 9 (1988), 387.
194. K. Balasubramanian and X. Liu, J. Comput. Chem. 9 (1988), 406.
195. E.C. Kirby, J. Math. Chem. 1 (1987), 175.
196. I. Gutman, Math. Chem. 23 (1988), 95.
197. R. Ramaraj and K. Balasubramanian, J. Comput. Chem. 6 (1985), 122.
198. D.M. Berman and K.W. Holladay, J. Math. Chem. 1 (1987), 405.
199. M. Randić, H. Hosoya, N. Ohkami and N. Trinajstić, J. Math. Chem. 1 (1987), 97.

200. I. Gutman, Publ. Inst. Math. (Beograd) **37** (1985), 29.

Chapter 5

ENUMERATION OF ISOMERS

Alexandru T. Balaban

Organic Chemistry Department, Polytechnic Institute,
Bucharest, Romania

5.1 Introduction

The phenomenon of isomerism is of fundamental importance in chemistry. It has dominated organic chemistry for over a century and also plays a role in many classes of inorganic compounds especially in organometallic complexes. Owing to the directional and localized character of simple covalent bonds, a collection of atoms may be linked in several nonisomorphic ways. This gives rise to the appearance of isomers.

Students of organic chemistry who are taught the theory of chemical structure soon acquire the facility of discovering by trial and error how many isomeric structures of molecules containing a few multivalent atoms can be drawn. This practice is actually an unconscious application of strict and rigorous mathematical rules. For larger numbers of atoms this "trial-and-error" method fails, however, as any trained chemist may verify by trying to write the 355 possible constitutionally isomeric dodecanes. The method of arriving at the number 355 will be mentioned again in the historical paragraph. When heteroatoms, rings, or unsaturation are involved, the problem becomes even more complicated because then one has to consider, for instance, the instability of enols, of saturated chains containing only heteroatoms, and of sterically strained systems.

The search for rigorous mathematical formulas or algorithms enabling chemists to cope with complicated problems during the second half of the last century contributed to the birth of the mathematical discipline known as graph theory (about which more will be said in subsequent paragraphs). In more recent times, with the advent of computers, the quest for good algorithms for isomer enumeration purposes has acquired a new dimensions; this problem has been selected as a promising demonstration of artificial intelligence [1,2].

Isomer enumeration methods have been reviewed several times. The literature till 1973 is summarized in Rouvray's excellent and comprehensive review [3] which emphasizes the classical methods for enumerating alkanes and alkyl derivatives; a similar approach is contained in Trinajstić's recent book [4]. In a book edited by the present author in 1976, three chapters were devoted to this problem: one dealing with Pólya's contributions to the field [5a], a second with the enumeration of acyclic systems [5b], and a third with the enumeration of cyclic systems [5c]. Since that time considerable progress has been made in the latter area. Accordingly, the present review will stress these more recent applications, while trying to reduce repetition of previously reviewed work to the minimum necessary to make the present chapter intelligible and self-contained.

Several other, less extensive, reviews on isomer enumeration have also appeared [6,7].

In passing, mention should be made of the many books on stereochemistry and stereoisomerism, among which only a few can be cited [8–19]. A number of authors have also treated the theoretical aspects of isomerism [19,20].

5.2 Definitions and Mathematical Background

5.2.1 Constitutional and Steric Isomerism

It is appropriate to start with the definition of isomerism. Different chemical substances (at least by some experimentally determinable properties) are said to be isomers if and only if they have the same molecular formula. This definition restricts the notion of isomerism to compounds containing covalent bonds (for only then can one speak of molecules and molecular formulas), and possibly also other kinds of bonds, such as ionic bonds. For example, urea and ammonium cyanate are isomers. Wöhler's isomerization of the latter into the former substance (1828) is considered to represent the "*experimentum crucis*" in disproving the vitalist theory. The above definition also implies that two isomeric molecules consist of the same set of atoms bonded differently.

Traditionally, isomerism has been classified as either constitutional or steric, with the latter class being subdivided into enantiomers and diastereoisomers. The flowchart, Figure 1, illustrates this traditional subdivision.

Constitutional isomerism is exemplified by ethanol and dimethyl ether (with differing heteroatom-containing chains C–C–O and C–O–C) or by butane and isobutane having linear or branched chains. The word "constitution" implies topological relationships, i.e. information on the neighborhoods of each atom. Constitutional formulas therefore contain only such information, and are thus equivalent to non-directed graphs (or multigraphs, if multiple bonds are present). Bond angles, bond lengths and distances between non-bonded atoms are irrelevant to constitutional formulas (which are sometimes called "line formulas" [21]).

Stereoisomerism takes into account, in addition to constitution, the above factors (bond lengths, bond angles, distances between non-bonded atoms — these parameters define diastereoisomerism) as well as the spatial arrangement of bonds around atoms i.e. the configuration (this

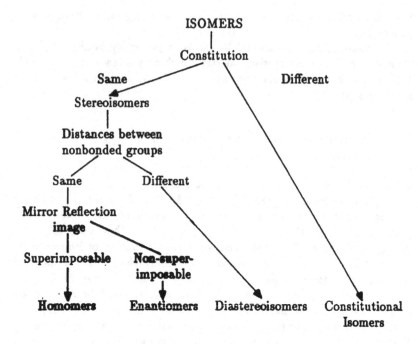

Figure 1

arrangement defines enantiomerism). A pair of enantiomers is related by mirror reflection. A chiral object is not superimposable on its mirror image and thus enantiomers are chiral.

An alternative subdivision was proposed by Mislow [25], who defined isometric and anisometric structures based on the isometry operations I_1 and I_2. An identity of all scalar properties is an isometric operation I_1; an operation of mirror reflection is an isometric operation I_2. If two structures can be derived from one another by one of these two operations, they are called isometric; otherwise they are called anisometric. Anisometric structures may be further divided according to their constitution. The resulting flowchart (Figure 2) leads once more to the same four categories. Homomers are structures which cannot be distinguished in chiral or achiral media; they may, however, differ by more subtle criteria such as isotopic composition. Enantiomers cannot be distinguished in achiral media but they can in chiral media. Diastereoisomers differ in their scalar properties in both chiral and achiral media though they have the same constitution. Constitutional isomers share neither the same constitution nor scalar properties, and are therefore easily distinguished in chiral or achiral media. Mislow's flowchart is a more equilibrated division than the more traditional flowchart.

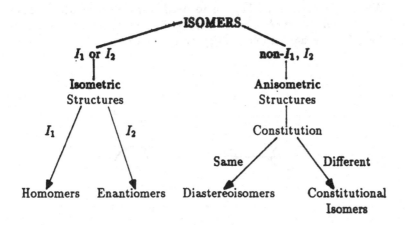

Figure 2. Mislow's classification for constitutional and steric isomerism.

5.2.2 Graph-Theoretical Background

A few graph-theoretical definitions will be necessary. We shall discuss here connected graphs, in which any point (vertex) can be reached from any other vertex via a path consisting of adjacent bonds (edges). Graphs having no cycles are called *trees*. If one point of a tree is distinguished from the others we have a rooted tree with the root at the respective point. Rooted trees may be used to symbolize alkyls, while unrooted trees symbolize alkanes, if the valence (vertex degree) is at most 4. Graphs having only single bonds are called *simple graphs*, whereas graphs with multiple bonds are called *multigraphs*. If the number of points (the "order" of the graph) is denoted by p, and the number of lines by q, then $\mu = q - p + 1$ is called the *cyclomatic number* of the graph; for trees having no cycles, $\mu = 0$ and $p = q + 1$.

The *adjacency matrix* of a graph is made up of the entries 1 for adjacent vertices (connected by an edge), and 0 otherwise. The sums of the entries over rows or columns in this matrix are graph invariants called *vertex degrees*: irrespective of the arbitrary labelling of vertices, the number of edges meeting at a vertex is fixed. Graphs having all vertices of the same degree are called *regular graphs*; in particular, regular graphs of degree three are called *cubic graphs*.

When a vertex can be bonded to itself (through a loop), we have the class of *general graphs*. In the degree the loop is counted twice, so therefore, 1 is a general cubic multigraph of order six.

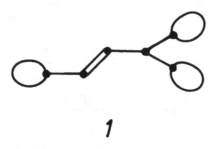

1

The topological distance (or more briefly the *distance*) between two vertices in a graph is the number of lines in the shortest path between these two vertices. Distances are finite numbers in all connected graphs. The *distance matrix* is a square $p \times p$ matrix whose entries are the distances between the edges. The adjacency and distance matrices are similar for entries 1, and both have 0's on the main diagonal, whereas all other off-diagonal 0 entries in the adjacency matrix are replaced by numbers larger than 1 in the distance matrix. Sums over rows or vertices in the distance matrix constitute another graph invariant, called *distance sums* (or distance degrees).

Labelling						*Vertex*
$\downarrow \rightarrow$	1	2	3	4	5	*degrees*
1	0	1	0	0	0	1
2	1	0	1	1	0	3
3	0	1	0	0	0	1
4	0	1	0	0	1	2
5	0	0	0	1	0	1

Adjacency matrix of 2

Labelling   *Distance*

↓→	1	2	3	4	5	*sums*
1	0	1	2	2	3	8
2	1	0	1	1	2	5
3	2	1	0	2	3	8
4	2	1	2	0	1	6
5	3	2	3	1	0	9

Distance matrix of 2

An example is provided by the graph 2 of isopentane.

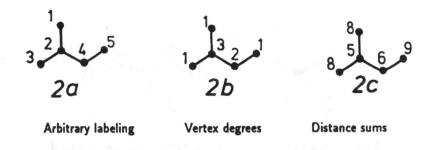

2a	*2b*	*2c*
Arbitrary labeling	Vertex degrees	Distance sums

Other graph-theoretical terms and definitions will be introduced below, and may also be found in books and reviews [23].

Graph centre, bicentre centroid and *bicentroid.* It is well-known [7] that any tree possesses a centre; this is either one vertex (when it is referred to as a centre) or a pair of adjacent vertices (when the pair is called a bicentre). To find the (bi)centre one prunes sequentially the vertices of degree one (endpoints) of the tree; alternatively, one considers the eccentricities of vertices defined as the maximum topological distance to an endpoint; (bi)centres have minimum eccentricities. Cayley used in this case the term (bi)centre of number.

A distinct alternative type of central vertex (vertices) called a centroid is obtained by considering the weight at each vertex (equal to the maximum number of lines on any branch starting from this vertex); the (bi)centroid of a tree is the vertex (or the pair of adjacent vertices) with minimum weight. An equivalent procedure is to consider the distance

Eccentricities and centre

Distance sums and centroid

Eccentricities and bicentre

Distance sums and bicentroid

sums for each vertex: the (bi)centre has minimum distance sums. Cayley used in this case the term (bi)centre of length.

Tree 3 has a centre and a bicentroid, whereas tree 4 has a centroid and a bicentre. In other cases, the (bi)centre coincides with the (bi)centroid.

5.3 Historical

The first discussion of different substances which contain the same atoms bonded in various manners was presented in 1811 by Gay-Lussac [24]. Experimental evidence was presented by Liebig after he introduced his precise analytical methods [25]; having discovered silver fulminate in 1823, Liebig showed that it had the same formula as silver cyanate which had been characterized earlier as a different compound by Wöhler. In 1825 Faraday prophesied that this phenomenon would become the rule rather than the exception [26]. Berzelius who had investigated tartaric and racemic acids (both isolated from wine), proposed in 1830 the words "isomerism" and "isomer" (Greek, isos = equal, meros = part) [27]. Many more examples of isomers were discovered in the following decades but since the notions of atom and molecule, and of ionic and covalent bonds, were still poorly understood, the theory could not advance. Structure theory originated from ideas put forward by Frankland, Gerhardt, Laurent, Butlerov [28], and, above all, by Kekulé [29]. At the first international Chemistry Congress organized by Kekulé in 1860 in Karlsruhe, Cannizzaro drew attention to Avogadro's work which had clarified the concepts of atom and molecule. The graphical notation system still in use in chemistry today resulted from chemical diagrams due to Kekulé [29], Couper [30] and Crum Brown [31]. Kekulé's *Lehrbuch der organischen Chemie* appeared in 1861 and contained the first systematic exposition of the theory of chemical structure [29], although present day chemical diagrams came to be used only a few years later. The discovery of the two predicted isomeric alcohols C_3H_7OH (Friedel prepared isopropanol) and of the four alcohols C_4H_9OH (Butlerov prepared t-butanol) served as brilliant and convincing arguments for the correctness of structure theory (initially, the term isomer embraced all compounds with the same empirical formula, including polymers, but later was substituted for the term metamer).

Only after Le Bel [32] and van't Hoff [33] advocated the three-dimensional tetrahedral arrangement of covalent bonds around carbon atoms, and after Werner had applied this idea to organometallic complexes [34], were the most common types of isomerism satisfactorily explained.

There still remained opponents (such as Kolbe) to the idea that chemical formulas represented reality, though when physical and quantum chemistry gave results in total agreement with empirically deduced structure theory, the remaining doubts vanished.

The isomerism found in sugars necessitated the concept of conformation. The advent of NMR and other techniques which were able to determine the properties of rotational or other stereochemically nonrigid molecules raised new problems concerning the definition and limitation of the concept of isomerism. The field of chemical isomerism is an active one, as proved by the several reviews on isomerism [3,6,20], the many recent books on stereochemistry [8–19], and by the award of Nobel prizes to Barton in 1969, and Prelog in 1975.

Turning now to the history of isomer enumeration, in 1857 the mathematician Cayley [35], stimulated by his friend Sylvester (a mathematician interested in chemistry), developed the theory of enumerating rooted trees. He found a recursive formula giving the numbers A_i of rooted trees with i points (corresponding to the alkyl radicals with i non-hydrogen atoms if one ignores the restriction on the vertex degree which sets an upper limit of four bonds per carbon atom) as the coefficient of x^i in the series:

$$1 + A_1 x + A_2 x^2 + A_3 x^3 + \ldots = (1-x)^{-A_1}(1-x^2)^{-A_2}(1-x^3)^{-A_3} \ldots$$

Later, in 1874, Cayley succeeded in enumerating unrooted trees, which (but for the analogous limit on the vertex degree) correspond to alkanes. In 1875 he published a second paper explicitly mentioning the organic chemical application [36], in which he considered the restriction that no vertex degree should be larger than four, and produced extensive tables of numbers of constitutional isomers for alkanes. Cayley's results were correct for alkanes with 1-11 carbon atoms, but erroneous for those having 12 and 13 carbon atoms (in the former case he found 357 isomers and in the latter 799 isomers); the numerical results for alkyl radicals were correct for alkyls with 1-12 carbon atoms, but incorrect for $C_{13}H_{27}$ (7638 isomers). Schiff (a chemist teaching in Italy) published immediately afterwards [37] a note based on a cumbersome procedure for counting alkane isomers, in which he arrived at the same erroneous number of dodecane isomers. Five years later, in 1880, Hermann [38] found the correct number of dodecane isomers, namely 355; however, as late as 1893 Tiemann was still reporting the erroneous number [39].

Later workers (Delaunoy [40], Losanitsch [41], Goldberg, Trautz, David – all mentioned by Rouvray [3]) attempted to find a formula for the number of alkanes, but made errors and produced methods which were not applicable beyond $n = 14$.

In 1931, a new approach was initiated by Henze and Blair at the University of Texas at Austin [42], on the basis of recursive formulas. They calculated the correct values for alkanes with $n = 12$ and 13 (355 and 802, respectively) and for alkyls with $n = 13$ (7639 constitutional isomers) but listed for $n = 19$ an erroneous value which was corrected by Perry [43] to 148,284 (instead of 147,284, which may have been a typographical error).

In 1911, Burnside formulated a lemma (previously known to Frobenius) which is useful in graphical enumeration [44], but whose application in this area came much later.

In 1927, a mathematician, Redfield, wrote the only paper he ever published [45], though the paper was completely overlooked because of its abstruse language. A paper published jointly by a chemist, Lunn, and a mathematician, Senior, made use of the mathematical theory of permutation groups [46]. These three papers foreshadowed the fundamental work of George Pólya, an Hungarian-born American mathematician, who in 1935–1937 produced a series of papers which are fundamental in both mathematics and chemistry [47]. Pólya's fundamental theorem will be discussed in detail in a later section.

More recently, Ruch (a theoretical chemist formerly at West Berlin's Free University) developed a different approach based on group theory (double coset formalism) which can be successfully employed for isomer counting [48]. This approach entails less labour than the expansions of lengthy polynomial required by Pólya's theorem, but at the same time allows less flexibility in terms of figure weighting.

Mathematicians (De Bruijn [49], as well as Harary and Palmer [50]) formulated the power group theorem which can be applied to the enumeration of isomers. Chemists have also contributed to isomer enumeration, either alone or in collaboration with mathematicians. Balaban [51] has provided a constructive algorithm for finding the structures and enumerating valence isomers of annulenes and of other systems, and Balaban and Harary [52–55] have proposed new definitions to assist in the enumeration of cata/peri-condensed benzenoid hydrocarbons [52], or counted isotope isomers [53], steric trees [54], monocyclic and polycyclic aromatic and heteroaromatic compounds [55]; a sizable number of Balaban's papers in the series "Chemical Graphs" deals with isomer enumeration. In the USA, Lederberg and his associates [1,2] have developed powerful computer programs for generating and counting isomers; several of their papers in the series "Applications of Artificial Intelligence for Chemical Inference" are devoted to this problem.

5.4 Pólya's Theorem

5.4.1 Enumeration of Alkane and Alkyl Constitutional Isomers

It was mentioned in our historical introduction that Cayley [35,36] counted first the rooted trees without restriction of vertex degree, and then went on to enumerate the chemically more interesting case when the vertex degree is at most four. Rooted trees correspond to the constitutional isomers of the alkyls (or alcohols, haloderivatives, etc.), whereas (unrooted) trees correspond to the constitutional isomers of the alkanes. Cayley discussed separately trees with a centre and bicentre and later used a simpler approach involving both centroids and bicentroids. Another relevant contribution of Cayley is the formula for counting labelled trees, i.e. trees where each vertex bears a label (chemically, this label may indicate the nature of the element or the isotope at the corresponding vertex). The number of labelled trees having n vertices (with no restriction on vertex degrees) is t_n, where

$$t_n = n^{n-2}.$$

Thus, there exist $4^{4-2} = 16$ labelled trees on four vertices: twelve are isomorphic to the skeletal graph of butane, and four to that of isobutane.

The Henze-Blair approach [42] is a recursive one; Trinajstić's book [4] exemplifies this approach for calculating the number (355) of constitutional isomers for dodecanes. A computer program based on this approach has been developed [56,57]. See also Section 5.12.

5.4.2 Pólya's Enumeration Method

Pólya's celebrated counting theorem (the Hauptsatz) [47] enumerates isomers or other "equivalence classes of functions" by taking into account the symmetry operations, namely the proper rotation axes, of the structures. The symmetry is expressed as the cycle index Z, i.e. a polynomial in terms of variables s_k which express the symmetry of the molecule. The various numbers of isomers are obtained as coefficients of the *configuration-counting series* by substituting into the expression for the cycle index Z a *figure-counting series* for each s_k.

As a first example, we shall discuss (Figure 3, Table 1) the benzene molecule (planar regular hexagon) which has 12 symmetry operations (expressed as cyclic permutations).

In order to apply Pólya's theorem, we substitute for hydrogen-labelled benzene (corresponding to substitution of one or more hydrogen atoms by a halogen atom or a hydrogen isotope) the figure counting series:

$$s_k = 1 + x^k$$

and thereby obtain the counting polynomial (configuration-counting

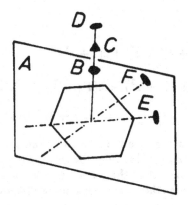

Figure 3. The Symmetry Axes of Benzene.

Table 1. Generation of the cycle index for the benzene molecule

Axis	Symmetry Operation	Permutation	Cycle index term
A	One identity E	(1)(2)(3)(4)(5)(6)	s_1^6
B	Two 6-fold rotations $\pm C_6$	(123456), (165432)	$2s_6$
C	Two 3-fold rotations $\pm C_3$	(153)(264), (135)(246)	$2s_3^2$
D	One vertical binary axis C_2	(14)(25)(36)	s_2^3
E	Three in-plane binary axes C_2	(16)(25)(34), (12)(36)(45), (14)(23)(56)	$3s_2^3$
F	Three in-plane binary axes C_2	(1)(4)(26)(35), (3)(6)(25)(24), (2)(5)(13)(46)	$3s_1^2 s_2^2$

Cycle index:

$$Z = (s_1^6 + 2s_6 + 2s_3^2 + 4s_2^3 + 3s_1^2 s_2^2)/12.$$

series):

$$Z = \frac{1}{12}\left[(1+x)^6 + 2(1+x^6) + 2(1+x^3)^2\right.$$
$$+ 4(1+x^2)^3 + 3(1+x)^2(1+x^2)^2\right]$$
$$= 1 + x + 3x^2 + 3x^3 + 3x^4 + x^5 + x^6.$$

The coefficients of x^6, x^5 and x indicate that there exist one hexa-, one penta- and one monosubstituted benzene; the coefficients of x^4, x^3, and x^2 express the fact that there are three isomers of tetra-, tri- and disubstituted benzenes (the latter being the well-known *ortho*, *meta* and *para*-isomers).

The same number of isomers are found for carbon-labelled benzene (isotopically-labelled benzene with one other isotope, e.g. ^{13}C).

For hydrogen-labelled benzenes with two types of halogen, the figure-counting series becomes:

$$s_k = 1 + x^k + y^k,$$

and we obtain the terms in the counting polynomial $3xy$, $3x^2$, $3y^2$, $6xy^2$, $11x^2y^2$, etc. whose coefficients indicate the numbers of isomers (e.g. six isomers for $C_6H_3FCl_2$, eleven isomers for $C_6H_2F_2Cl_2$).

For carbon-labelled benzene with two other isotopes (e.g. ^{13}C and ^{14}C, in addition to the natural ^{12}C isotope) we obtain the same number of isotopes.

For carbon- and hydrogen-labelled benzene, we have to employ a different figure-counting series, since one and the same position can be doubly labelled thus:

$$s_k = 1 + x^k + y^k + x^k y^k.$$

We then obtain the following terms in the counting polynomial: $3x^2$, $4xy$, $9xy^2$, $12xy^3$, $24x^2y^2$, etc. The coefficients indicate [55] that there are four isomers of $^{12}C_5{}^{13}CH_5F$, nine isomers of $[^{13}C_1]$ difluorobenzene $^{12}C_5{}^{13}CH_4F_2$, or of $[^{13}C_2]$ fluorobenzene $^{12}C_4{}^{13}C_2H_5F$, twenty-four isomers of $[^{13}C_2]$ difluorobenzene $^{12}C_4{}^{13}C_2H_4F_2$, etc. (see also Table 5 in Section 5.5).

A related counting problem, known in graph theory as the "necklace counting problem" (since it is equivalent to counting nonisomorphic necklaces with beads of different colour), is encountered in the enumeration of all possible monocyclic aromatic systems possessing heteroatoms [58]. According to Pauli's principle, the atoms in an aromatic system can be of only three types, namely X, Y and Z with two, one and zero pi-electrons in the nonhybridized p_z-orbital. Six-membered heterocycles with six pi-electrons can therefore have formulas Y_6, XY_5Z, $X_2Y_2Z_2$, X_3Z_3. The numbers of isomers for these four types of heterocycles can be calculated using Pólya's theorem [52] to be 1,3,11, and 3, respectively (see the benzenes with one or two types of halogen). A direct formula solving this problem was published by Balaban and Teleman [58].

Since it is known that adjacent X-type, or Z-type, heteroatoms decrease the stability of species, it was interesting to enumerate cyclic systems where no such adjacencies exist; this was done by Lloyd [59].

Heterocyclic systems having chains of odd-numbered Y-type atoms separated by X and/or Z-type atoms are called "mesoionic"; this is an alternative definition to the usual one, first devised for sydnones. So far the only known mesoionic systems are five-membered ones, but the tabulated systems of X, Y, and Z-type heterocyclic mesoionic (mesoaromatic) systems satisfying Hückel's rule can also have larger numbers of atoms in the ring.

A further example of the usefulness of Pólya's theorem is provided by the toluene molecule. The cycle index for carbon-labelled toluene is:

$$Z(\text{toluene}) = (s_1^7 + s_1^3 s_2^2)/2$$

because in addition to the identity operation (s_1^7), there is a binary axis permuting two pairs of atoms which passes through three atoms ($s_1^3 s_2^2$). By substituting the figure-counting series $s_k = 1 \times x^k$, one obtains the configuration-counting series for carbon-labelled toluene:

$$x^7 + 5x^6 + 16x^5 + 21x^4 + 21x^3 + 16x^2 + 5x + 1$$

which indicates that there are five isomers of $[^{13}C_1]$ toluene, sixteen of $[^{13}C_2]$ toluene, etc.

For hydrogen-labelled toluene one has to consider the free rotation of the methyl group attached to the phenyl group. Character tables give the expressions for dihedral, symmetric, cyclic groups, etc; operations with these groups lead to the desired cycle indices. The product of the symmetry group

$$Z(s_3) = (s_1^3 + 3s_1 s_2 + 2s_3)/6$$

with the group corresponding to the five phenyl hydrogens

$$Z(B) = (s_1^5 + s_1 s_2^2)/2$$

is:

$$Z(B)Z(s_3) = \frac{1}{12}(s_1^8 + 3s_1^6 s_2 + 2s_1^5 s_3 + s_1^4 s_2^2 + 3s_1^2 s_2^2 + 2s_1 s_2^2 s_3).$$

If one now substitutes the figure-counting series $s_k = 1 + x^k$, one obtains the counting polynomial:

$$x^8 + 4x^7 + 110x^6 + 16x^5 + 18x^4 + 16x^3 + 10x^2 + 4x + 1,$$

indicating that there are, for instance, four isomers of the monochloro-substitution product of toluene, namely benzyl chloride and the three o, m, p-chlorotoluenes [54].

Pólya's theorem can be applied to a large variety of enumeration problems [60–70]. Certain problems solved by cumbersome methods [70] could have been readily solved using this theorem. Mathematicians have also used Pólya's theorem to show its potential applications in chemistry [5a,5b,71,72].

The correct numbers of substituted porphyrins were found [73] by applying Pólya's theorem; thereby a correction to an alleged correction [74] of a textbook error was made (corrections using less elegant methods were also published [75,76]).

The next example is of the isomers for substituted cycloalkanes [77], assumed to undergo such rapid configurational changes as to average the environment of equatorial/axial substituents, so that only the average planar ring conformation need be considered [78]. We shall illustrate the enumeration with substituted cyclopropanes whose cycle indices can be obtained from the symmetries of the graphs 5–7.

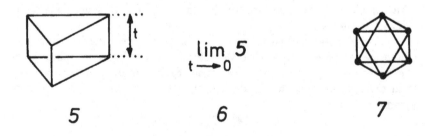

5   6   7

For all isomers, including stereoisomers, one has to consider the prism-like geometry 5; for the omission of enantiomers so that only constitutional and geometrical (*cis-trans*, or *E–Z*) isomers are counted, one has to consider graph 6 which is the limit of the prism when its thickness becomes zero. Finally, for counting only constitutional isomers, non-planar graphs of degree four (7) have to be considered. For an m-membered cycloalkane, such graphs (called Balaban graphs by Mallion [79] consist in a $2m$-membered ring with two "internal edges" connecting vertex i to the two vertices adjacent to the vertex opposite i, namely the vertices numbered $i + m + 1$ and $i + m - 1$. The result for cyclopropane is given in Table 2.

Table 2. Cycle indices and counting polynomials
for substituted cyclopropane.

$$Z(5) = \frac{1}{6}(s_1^6 + 2s_3^2 + 3s_2^3)$$

$$Z(5, 1 + x) = 1 + x + 4x^2 + 4x^3 + 4x^4 + x^5 + x^6$$

$$Z(6) = \frac{1}{12}(s_1^6 + 2s_3^2 + 4s_2^3 + 3s_1^2 s_2^2 + 2s_6)$$

$$Z(6, 1 + x) = 1 + x + 3x^2 + 3x^3 + 3x^4 + x^5 + x^6$$

$$Z(7) = \frac{1}{48}(s_1^6 + 8s_3^2 + 7s_2^3 + 9s_1^2 s_2^2 + 8s_6 + 3s_1^4 s_2 + 6s_1^2 s_4)$$

$$Z(7, 1 + x) = 1 + x + 2x^2 + 2x^3 + 2x^4 + x^5 + x^6$$

Thus, there are four A_2-disubstituted isomers 8–11 if all types of stereoisomerism are considered. If enantiomerism is ignored, there remain only three isomers: 8, *cis* (9) and *trans*, since 10 and 11 are the pair of enantiomeric *trans*-1,2-disubstituted cyclopropanes. If only constitutional isomerism is considered, there remain two constitutional isomers, namely 8 (1,1-disubstituted) and 9–11 (1,2-disubstituted). This discussion illustrates the power of Pólya's method, provided the appropriate graph is employed (which may differ from the intuitive graph, as indicated by graph 7).

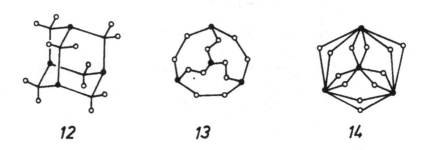

8   9   10   11

For the related problem of finding the number of chiral 1,2,3,4,5,6-hexachlorocycloalkanes, Dias [80] described a procedure based on Pólya's theorem whereby this number is $Z(C_n, 1 + x) - Z(D_n, 1 + x)$, leading to one pair of enantiomers for rings with $n = 6$ and 7 atoms, and four pairs for $n = 8$.

Enumerations of substituted adamantanes using Pólya's theorem were carried out by Balaban, Palmer and Harary [81], and generalized to polymantane structures by Balaban and Baciu [82]. Again, in order to count the stereoisomers, geometrical isomers or constitutional isomers separately, one has to depart from the normal graph of adamantane (**12**), whose symmetries allow the counting of all stereoisomers, and use graph **13** for counting all the geometrical and constitutional isomers (ignoring enantiomerism), or graph **14** for counting the constitutional isomers only.

12   **13**   **14**

The numbers of stereoisomers of adamantane derivatives were also investigated by Masinter *et al* [83], by Kornilov [84] (who used Pólya's theorem) and by Ruch *et al* (who used the double coset formalism) [48].

Counting the constitutional isomers of trees, even with the restriction that no vertex should have degree higher than four, is a problem which was solved long ago by Cayley [35,36], followed by Henze and Blair [42], and then Pólya [47]. However, alkanes can be chiral if they have at least seven carbon atoms. Counting stereoisomeric alkanes is a more difficult problem, tackled by Read, who applied Pólya's theorem for stereo-hydrocarbons of acyclic hydrocarbons with simple, double and/or triple bonds [56,72].

Robinson, Harary and Balaban [54], by applying Pólya's theorem and Otter's dissimilarity characteristic equation [85], presented for the first time recurrence formulas for counting the achiral isomers of alkyl radicals and alkanes. If the constitutional isomers of alkanes are denoted by v_n (quartic trees), and if their number, including stereoisomers, is denoted by f_n (steric trees), the latter number may be decomposed into achiral (t_n) and chiral (e_n) isomers. The numbers for $n = 1\text{--}14$ are presented in Table 3;

$$v_n = t_n + e_n/2; \qquad f_n = t_n + e_n.$$

Similar relationships hold for the monosubstituted alkanes (planted trees) and are also presented in Table 3.

Table 3. Number of alkanes and monosubstituted alkanes

n	Alkanes				Substituted alkanes	
	Achiral t_n	Chiral e_n	Steric f_n	Quartic v_n	Steric f_n'	Quartic v_n'
1	1	0	1	1	1	1
2	1	0	1	1	1	1
3	1	0	1	1	2	2
4	2	0	2	2	5	4
5	3	0	3	3	11	8
6	5	0	5	5	28	17
7	7	4	11	9	74	39
8	14	10	24	18	199	89
9	21	34	55	35	551	211
10	40	96	136	75	1553	507
11	61	284	345	159	4436	1238
12	118	782	900	355*	12832	3057
13	186	4226	2412	802	37496	7639
14	365	6918	6563	1858	110500	19241

*See the discussion in the historical section

Many more applications of Pólya's theorem are known, and some are presented in chapters 2 and 3 of a book cited above [5]. Pólya's Hauptsatz is based on Burnside's Lemma [44], but chemists unaware of this derivation could rediscover the latter Lemma starting from Pólya's theorem [67].

We conclude this section by showing how one can enumerate the constitutional isomers of the alkylbenzenes, C_nH_{2n-6}, with one benzene ring and $n \geq 7$. Upon denoting the number of different alkyl radicals having i carbon atoms by A_i, the configuration-counting series becomes:

$$Z(D_6, A(x)).$$

We know the expression of $Z(D_6)$ and we have the figure-counting series for alkyl radicals:

$$A(x) = A_0 + A_1x + A_2x^2 + A_3x^3 + \dots .$$

We can therefore obtain the counting series as follows:

$$\frac{1}{12}\left[A^6(x) + 4A^3(x^2) + 2A^2(x^3) + 3A^2(x)A^2(x^2) + 2A(x^6)\right].$$

The results for $n = 7$–17 are shown in Table 4. For instance, the 22 alkylbenzenes isomer counts with $n = 10$ are four $C_6H_5 - C_4H_9$, three $Me - C_6H_4 - nPr$, three $Me - C_6H_4 - iPr$, three $Et_2C_6H_4$, six $C_6H_3Me_2Et$ and three $C_6H_2Me_4$; one isomer, $sBuPh$, is chiral (this is not considered in the above enumeration).

Table 4. Isomers of the alkylbenzenes, C_nH_{2n-6}

n	7	8	9	10	11	12	13	14	15	16	17
Isomers	1	4	8	22	51	136	335	871	2217	5749	14837

5.5 Generalized Pólya Theorem

When there exist vertices symbolizing identical but chemically different atoms, or chemically non-equivalent atoms, it is advantageous to appropriately segregate these vertices from the start.

A first example is C/H-labelled ethene which for H-labelling has the configuration-counting series:

$$1 + h + 3h^2 + h^3 + h^4$$

and for C-labelling the configuration-counting series:

$$1 + c + c^2$$

as illustrated by Figure 4.

The symmetry operations are:

E	(1)(2)(3)(4)(5)(6)	$s_1^4 t_1^2$
C_{2x}	(12)(34) (5)(6)	$s_2^2 t_1^2$
C_{2y}	(14)(23) (56)	$s_2^2 t_2$
C_{2z}	(13)(24) (56)	$s_2^2 t_2$

Therefore, the cycle index assumes the form:

$$Z(D_2;\ s,t) = (s_1^4 t_1^2 + s_2^2 t_1^2 + 2s_2^2 t_2)/4.$$

By substituting two figure counting series $s = 1 + h$, $t = 1 + c$ into this cycle index we obtain the configuration-counting series:

$$S = (1 + c^2)(1 + h + 3h^2 + h^3 + h^4) + c(1 + 2h + 4h^2 + 2h^3 + h^4)$$

in agreement with experimental results.

A second example is provided by the trigonal bipyramid 15:

$$Z(D_{3h};\ s,t,u) = u(s_1^3 t_1^2 + 2s_3 t_1^2 + 3s_1 s_2 t_2 + s_1^3 t_2 + 2s_3 t_2 + 3s_1 s_1 t_1^2)/12.$$

H- labelled   ethene

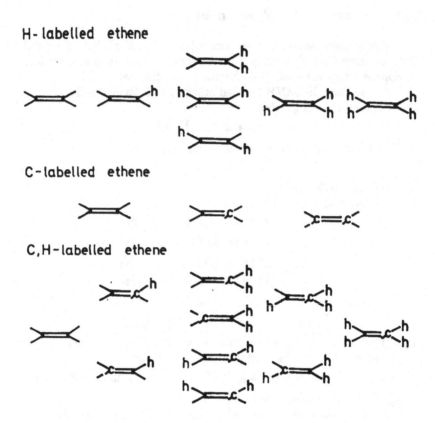

C-labelled   ethene

C,H-labelled   ethene

Figure 4. H-labelled, C-labelled and C,H-labelled ethene isomers

By substituting appropriate figure-counting series, one can calculate the number of isomers for apically ($t = 1 + a$), equatorially ($s = 1 + e$) or centrally ($u = 1 + c$) labelled isomers.

The third and last example is C/H labelled benzene. An alternative approach to the method described previously, in which the figure counting series

$$s_k = 1 + x^k + y^k + x^k y^k$$

was used, is to start from a cycle index having variable s for carbon symmetries and variable t for hydrogen symmetries:

$$Z(D_6;\ s,t) = (s_1^6 t_1^6 + 2s_6 t_6 + 2s_3^2 t_3^2 + 4s_2^3 t_2^3 + 3s_1^2 s_2^2 t_1^2 t_2^2)/12$$

and then substituting two figure-counting series, namely $s_k = 1 + c^k$, $t_k = 1 + h^k$. One thereby finds the numbers of isomers for the various partitions of the C/H labels indicated in Table 5.

Table 5. Numbers of isomers of C/H labelled benzene.

C or H	6	5+1	4+2	3+3	5+1	4+2	3+3	4+2	3+3	3+3
H or C	6	6	6	6	5+1	5+1	5+1	4+2	4+2	3+3
Isomers	1	1	3	3	4	9	12	24	28	38

The magnetic non-equivalence of diastereotopic groups by internal comparison is detectable from the number of signals in NMR spectra. To account for this number, Balasubramanian [86], applied a procedure patterned after Pólya's theorem, but making use of wreath products. Thus, for $CH_3 - CHCl - CH_2Cl$ it is found that the number of chemical-shift non-equivalent protons is four, since the coefficient of $y_1^{m-1} y_2$ in the figure counting series is 4. With six protons, the cycle index of the wreath product is

$$(s_1^6 + 2s_3 s_1^3)/3$$

and the counting series is

$$\left[\left(\sum_{i=1}^{6} f_i\right)^6 + 2\left(\sum_{i=1}^{6} f_i^3\right)\left(\sum_{i=1}^{6} f_i\right)^3\right]/3.$$

The coefficient of $f_1^5 f_2$ in the above series is indeed found to be

$$\frac{1}{3}\left[\binom{6}{5} + 2\binom{3}{2}\right] = 4.$$

The same approach explains why all six protons of benzene are equivalent (homotopic) [87].

Similarly, by using wreath products, Balasubramanian calculated other numbers of isomers such as atropisomers of the p-polyphenyls, the stereoposition isomers of freely rotating alkyl halides, etc. [86].

5.6 Ruch's Double Coset Formalism

In applying Pólya's enumeration theorem, the algebraic calculations upon substituting the figure-counting series can become very cumbersome; the information provided by the full configuration series is often redundant. Though sometimes we only need one coefficient, we have to carry out operations which automatically lead to all the coefficients. In many such applications the very breadth of Pólya's formulation makes it less attractive than Ruch's double coset formalism [48], which entails less labour than lengthy polynomial expansions.

We postulate a frame of sites $1, 2, \ldots n$ subject to an equivalence group G; these are segregated by type: n_1 of type 1, n_2 of type 2, etc., where $\sum_i n_i = n$. The indices of equivalent figures are symmetrized by a figure partition group $P = S_{n_1} \times S_{n_2} \times \ldots = \prod_i S_{n_i}$. On this basis, ordered configurations are bijections between the set $1, 2, \ldots n$ of site coordinates s and the set $1, 2, \ldots n$ of figure coordinates f represented by $2 \times n$ matrices

$$M_i = \left\| \begin{matrix} s \\ f_i \end{matrix} \right\| = \left\| \begin{matrix} 1 & 2 & 3 & \ldots & n \\ i_1 & i_2 & i_3 & \ldots & i_n \end{matrix} \right\|.$$

In order to obtain the desired equivalence classes among the $n!$ mappings thus defined, one associates configurations with permutations of the symmetric group $S_n = \{s_1, s_2, \ldots, s_{n!}\}$. The frame group G is a subgroup of S_n. Configurations equivalent to s_i due to frame symmetry constitute the left coset of G with respect to s_i, $s_i G$.

Configurations equivalent to s_i under figure index symmetry constitute the right coset of the isomorphic image of P with respect to s_i, $P s_i$.

For both the frame and figure symmetries, equivalent configurations comprise the double coset $P s_i G$ of P and G with respect to s_i. Thus, the symmetric group S_n is decomposed into double coset classes: $S_n = \bigcup P s_i G$. The number of double coset classes in S_n relative to subgroups P and G is according to Ruch:

$$d = \frac{|S_n|}{|G|\,|P|} \sum_n \frac{|G \cap C_r| \cdot |P \cap C_r|}{|C_r|}$$

where

$$C_r = \frac{n!}{\prod (k^{j_k} j_k!)}$$

is Cauchy's cardinality expression for the conjugacy class $(1^{j_1} 2^{j_2} \ldots n^{j_n})$ and $G \cap C_r$ or $P \cap C_r$ are, respectively, the number of permutations in G and P with this cycle structure.

On the allene frame $1 \ldots 4$, for instance, we have four ligands which may be partitioned as follows: $4, 3+1, 2+2, 2+1+1, 1+1+1+1$. As a first example we shall discuss the partition $2+1+1$. The number of constitutional isomers is:

$$S' = \frac{4!}{8\,2}\left(\frac{1\,1}{1} + \frac{2\,1}{6}\right) = 2$$

whereas the number of stereoisomers is:

$$S' = \frac{4!}{4\,2}\left(\frac{1\,1}{1}\right) = 3.$$

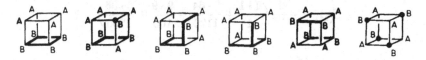

In the following example we shall use for partitions a more compact notation, e.g. instead of $4+4$ we shall write (4^2), and instead of $1+1+3+3$ we shall write $(1^2, 3^2)$, etc. i.e. k^{j_k} (Table 6).

Figure 5 displays the isomers of tetrasubstituted cubane.

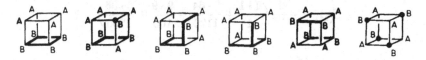

Figure 5. Isomers of tetrasubstituted cubane.

Table 6. The 4^2 Configurations of naphthalene and cubane

Partition	(1^8)	$(1^4,2^2)$	$(1^2,3^2)$	(2^4)	(4^2)	$(2,6)$	Observation
$P_{(4^2)} = S_4^2$	1	42	64	9	36	–	
$\cdot\ S_8$	1	210	1120	105	1260	3360	
Naphthalene $|G_{S=S'}| = 4$	1	–	–	3	–	–	Constitutional
Cubane $|G_S| = 24$	1	–	8	9	6	–	Stereo
$|G_{S'}| = 48$	1	6	8	13	12	8	Constitutional

The number of isomers for the partition $4+4$ is:

$$\text{Naphthalene} \qquad S_{(4^2)} = \frac{8!}{4(4!)2}\left(\frac{1\cdot 1}{1} + \frac{3\cdot 9}{105}\right) = 22.$$

These 22 isomers are indicated below by considering A equal to hydrogen and by listing the positions of the B substituents: 1234, 1235, 1236, 1237, 1238, 1245, 1246, 1247, 1248, 1256, 1257, 1258, 1267, 1268, 1278, 1357, 1358, 1367, 1368, 1467, 1468, 2367.

$$\text{Cubane} \qquad \text{(constitutional isomers)}$$

$$S'_{(4^2)} = \frac{8!}{2(4!)^3}\left(\frac{1\cdot 1}{1} + \frac{6\cdot 42}{210} + \frac{8\cdot 64}{1120} + \frac{13\cdot 9}{105} + \frac{12\cdot 36}{1260}\right) = 6$$

$$\text{Cubane} \qquad \text{(stereoisomers)}$$

$$S_{(4^2)} = \frac{8!}{(4!)^3}\left(\frac{1\cdot 1}{1} + \frac{8\cdot 64}{1120} + \frac{9\cdot 9}{105} + \frac{6\cdot 36}{1260}\right) = 7$$

5.7 De Bruijn-Harary-Palmer Power Group Theory

Applying Burnside's lemma [44] to the Harary-Palmer [50] power groups H^G, De Bruijn [49] developed an approach whereby one can calculate isomer numbers. Without entering into details, which can be found in references [20,49], we indicate that the necklace problem, i.e. H-labelled benzenes or azabenzenes, can be solved more directly by this method than by using Pólya's enumeration theorem. Particularly interesting stereoconfigurations appear when a regular frame, e.g. a hexagon, is labelled with substituents which may be identical or enantiomeric. By comparing the Pólya, Ruch and De Bruijn-Harary-Palmer approaches, the latter two methods afford more easily the correct enumerations.

5.8 Valence Isomers

Chemists have observed that some compounds react as if they possessed two rapidly equilibrating structures which differ only by the arrangement of the σ or π electrons. Such an example is cyclooctatetraene 15 which can be hydrogenated to cyclooctane, but which, being devoid of conjugation (because of its tub-shaped geometry), cannot undergo Diels-Alder cycloaddition. Its valence isomer, bicyclo-[420]-octatriene 16 possesses a conjugated diene system which reacts normally with dienophiles. The activation parameters for the interconversion 15→16 are 28.1 kcal/mol and +1 cal/mol grad; the reverse reaction 16→15 has an activation energy of 18.7 kcal/mol [88,89]. The low activation energies for the thermal interconversion have been rationalized by Woodward and Hoffmann [90].

15   16

Since at room temperature the equilibration 15⇌16 is very fast, one may speak in this case of valence tautomerism (recalling the fast tautomeric equilibration which involves changing the position of a proton and of a corresponding pair of π-electrons).

A rigorous graph-theoretical definition of valence isomers considers these isomers as being represented by connected skeleton graphs having the same vertex degree partitioning. For instance, the eleven graphs in Table 7 represent [91] all the possible isomers of C_4H_4. These isomers can be grouped into four classes of valence isomers, according to the partitioning of those vertex degrees for vertices representing atoms of valence higher than one in the skeleton graphs (in this case the carbon atoms). It can be seen that there exist two, and only two, connected graphs for $(CH)_4$: these represent cyclobutadiene, 17, and its valence isomer, tetrahedrane, 18. Maier succeeded in preparing stable tetra-*t*-butyl-substituted derivatives of both 17 and 18 [92].

Table 7. Isomers of C_4H_4 Grouped into classes of valence isomers.

Vertex degree	Formula	Skeleton graphs			
$12 = 3 + 3 + 3 + 3$	$(CH)_4$	17	18		
$12 = 4 + 3 + 3 + 2$	$C(CH)_2CH_2$				
$12 = 4 + 4 + 2 + 2$	$C_2(CH_2)_2$				
$12 = 4 + 4 + 3 + 1$	$C_2(CH)(CH_3)$				

It may be observed that in all cases treated in Table 7, the sum of vertex degrees is constant (12). It should also be observed that disconnected graphs (two separate molecules of acetylene) do not qualify as valence isomers.

The problem of finding an algorithm for constructing all the valence isomers of the annulenes $(CH)_{2k}$ ($k = 1, 2, ...$) is equivalent to constructing all possible connected cubic graphs [51]. This was an unsolved problem in graph theory in the early 1960s until Balaban solved it by using a recursive construction algorithm involving general graphs [93].

For $(CH)_6$ there exist five possible planar cubic graphs 19–23 and one nonplanar graph 24 [51,93]. In graph-theoretical language, *planar graphs* can be represented on paper without crossing any lines, whereas nonplanar graphs cannot. Planar graphs need not correspond to planar molecules. Indeed, though 19–23 have planar graphs, only benzene 19 is a planar molecule. All the other molecules 20–23 are tridimensional; only 24 has a nonplanar graph. Till now, no molecule is known whose covalent bonds form the edges of a nonplanar graph. However, if an edge of a nonplanar graph is allowed to correspond to a sequence of several covalent bonds, the synthesis of such molecules has recently been reported [94].

The five valence isomers 19–23 are all known: benzene, 19, Dewar benzene, 20, bicyclopropenyl, 21, benvalene, 22, and benzprismane, 23. The total number of graphs for C_6H_6 is 217, and, as for cyclobutadiene isomers, these may be grouped into seven classes of valence isomers: $6(CH)_6$, $32C(CH)_4CH_2$, $76C_2(CH)_2(CH_2)_2$, $16C_3(CH_2)_3$, $7C_4(CH_3)_2$, $34C_2(CH)_3(CH_3)$ and $46C_3(CH)(CH_2)(CH_3)$.

All these structures were displayed for the first time by Balaban [91], and have since been used in advertisements of the Jeol Company [95a] and on the cover of the magazine Chemie in unserer Zeit [95b].

For $(CH)_8$ there exist 17 possible planar cubic graphs, of which 12 are known as chemical compounds, examples being 15 and 16. Semibullvalene, 25 undergoes the most rapid degenerate valence isomerization among all known valence isomers of annulenes. For $(CH)_{10}$ there exist 71 possible planar cubic graphs [51,96], 26 of which are known as chemical compounds; Nenitzescu's hydrocarbon, 26, is the first valence isomer of any annulene to have been prepared [97a]. Bullvalene, 27, is one of the most interesting susbtances because by degenerate thermally-allowed Cope rearrangements at room temperature, all the CH groups appear equivalent in the ^{1}H-NMR spectrum. Pentaprismane, 28, was recently prepared [97b]. For $(CH)_{12}$ there exist 506 possible planar cubic graphs [98], only 10 of which are known as chemical susbtances; dodecahedrane 29 was recently prepared by Paquette [97c].

A formula published by Harary and Palmer [50] allows the counting of general cubic graphs on $2k$ vertices (connected and disconnected):

$$Z(S_{2k}[S_3]) \cap Z(S_{3k}[S_2])$$

where $\cap$ denotes the "cap" operation in Redfield's Theorem [45]. For

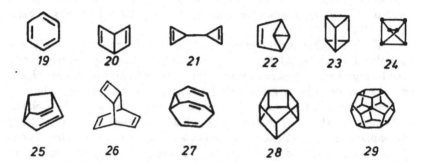

practical implementation, this formula needs a computer program which we are developing at present [99]. The number of disconnected general graphs can be calculated by a recursive formula, which enables us to find the number of connected general cubic graphs [99] by subtracting.

	General cubic graphs			Connected multigraphs		
n	Disconnected	Connected	Total	Planar	Non-planar	Total
4	3	5	8	2	0	2
6	14	17	31	5	1	6
8	69	71	140	17	3	20
10	334	388	722	71	20	91

In accord with Balaban's values for the valence isomers of [2k] annulenes [51,93] both mathematicians [100a] and chemists [100b] have found consistent numerical values for cubic graphs and multigraphs.

Another important class of hydrocarbons, in addition to the annulenes and their valence isomers, is represented by the benzoannulenes. Thus, naphthalene can be considered to be benzo-[6]-annulene. Benzoderivatives of the valence isomers of the [n]-annulenes were enumerated by Balaban [101] for $n = 4, 6, 8,$ and 10. Banciu and Balaban [102] summarized the results and indicated schemes for interconversions between these valence isomers.

According to the definition, one has to draw a distinction between the benzoderivatives of the valence isomers of annulenes (a restricted class), and the valence isomers of benzoannulenes (a much larger class). The difference arises because in the former case m pairs of vertices of degree 4 must be adjacent for the m-benzoderivatives of the valence isomers of [n]-annulene, e.g. one pair for the monobenzoderivative, while in the latter case no such adjacency restriction exists. For $n = 4, 6, 8$, and 10, the planar connected valence isomers of the benzoannulenes were investigated by Banciu and Balaban. For instance, monobenzo-[8]-annulene has 21 valence isomers in the former case and 355 valence isomers in the latter case. The valence isomers of the heteroannulenes [104] compound the problems of enumerating the cubic and rooted graphs corresponding to the valence isomers. (Rooted graphs are used if only one heteroatom is involved, e.g. for the enumeration of all the valence isomers of pyridine, i.e. of monoaza-[6]-annulene). This is done most simply by applying Pólya's theorem to cubic multigraphs [104]. The result for the valence isomers of benzene is presented in Table 8 [104].

Table 8. Numbers of valence isomers of substituted [6]annulene.

Isomer	Cycle Index	Partition of Substituents					
		5,1	4,4	3,3	4,1,1	3,2,1	2,2,2
20	$(s_1^6 + 2s_2^3 + s_1^2 s_2^2)/4$	2	6	6	8	16	26
21	$(s_1^6 + 2s_1^4 s_2 + 2s_2^3 + s_1^2 s_2^2 + \\ +2s_2 s_4)/8$	2	5	5	7	12	20
22	$s_1^6 + s_1^4 s_2 + s_2^3 + s_1^2 s_2^2)/4$	3	7	8	11	20	29
23^a	$(s_1^6 + 3s_2^3 + 2s_3^2)/6$	1	4	4	5	10	18
19 and 23^b	$(s_1^6 + 4s_2^3 + 3s_1^2 s_2^2 + 2s_3^2 + \\ +2s_6)/12$	1	3	3	3	6	11

^{a}Including stereoisomerism; ^{b}Excluding stereoisomerism.

A few comments are necessary on the data in Table 8. The number of aza-valence isomers is the same as the number of substituted derivatives, e.g. there exist one monoazabenzene (pyridine, corresponding to partition 5,1 for 19), three diazines (partition 4,4 for 19), etc. (For a computer program generating all aza-benzenoids with up to ten benzenoid rings and up to eight nitrogen atoms, see Section 12. For benzvalene, 22, there exist three monoaza-, seven diaza-derivatives, etc. Interestingly, the cycle index (hence the number of isomers for substituted derivatives) is the same for benzene, 19, and for benzprismane, 23, if one ignores stereoisomerism.

Such systems having the same cycle index are called *coisomeric graphs*. Historically, the fact that benzene and benzprismane are coisomeric was very important because it gave rise to a long polemic between Kekulé and Ladenburg [105a] who advocated the benzprismane formula for benzene. Indeed, as shown in Figure 6, substitution of disubstituted systems [105b] leads both for benzene and benzprismane to similar but different (for each type of disubstituted starting material) numbers of trisubstituted isomers.

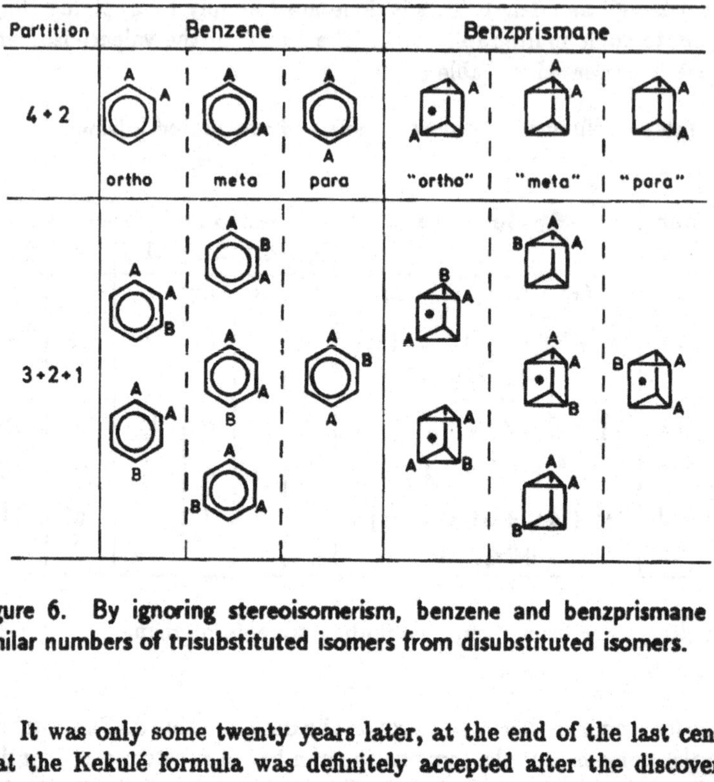

Figure 6. By ignoring stereoisomerism, benzene and benzprismane yield similar numbers of trisubstituted isomers from disubstituted isomers.

It was only some twenty years later, at the end of the last century, that the Kekulé formula was definitely accepted after the discovery of *ortho*-fused systems such as naphthalene (whose prismane formula was extremely awkward), and the failure to find any evidence for optical activity in systems such as those marked with an asterisk in Figure 6, which should be chiral.

Many more coisomeric valence isomers of the annulenes exist, such as the three pairs for $(CH)_6$ [104]: 30 with 31, 32 with 33, and 34 with semibullvalene 25.

30   31   32   33   34   25

The enumeration of homoannulenes is interesting because some of these systems are aromatic or homoaromatic. By definition, the skeleton graphs of homo-[n]-annulenes have in addition to the n vertices of degree 3, one vertex of degree 2 (for monohomosystems) or several such vertices (for polyhomosystems). The monohomo-[n]-annulenes can correspond to aromatic systems if $n = 4k$ and if the vertex of degree two is a heteroatom such as O, S, NR, etc.; they can correspond to homoaromatic systems if the vertex of degree two is a CH_2 group and if $n = 4k + 2$, though normally homoaromaticity is manifested in ions rather than in neutral systems.

For enumerating the monohomo-[n]-annulenes, the simplest approach was found to be an algorithm [93,106] which constructs general cubic graphs with $n+2$ vertices. From these, the general cubic graphs with one loop are selected; the loop and its attached edge are deleted, resulting in a cubic graph with n vertices of degree 3 and one vertex of degree 2. Thus, four general cubic graphs with one loop and six vertices exist (35–38); they correspond to the valence isomers of furan or thiophene ($X = O$ or S). Benzohomoannulene valence isomers were also enumerated by a similar algorithm [107].

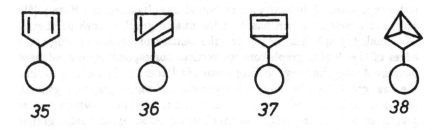

35   36   37   38

The valence isomers of trees were investigated by Quintas and coworkers [108], starting from relationships between the numbers of vertices and edges, respectively:

$$p + s + t + q = n$$
$$p + 2s + 3t + 4q = 2(n - 1)$$

where the number n of vertices is partitioned according to the degrees 1, 2, 3, and 4 into primary (p), secondary (s), ternary (t) and quaternary (q) ones. By taking t and q as independent parameters, one obtains

$$p = 2 + t + 2q$$
$$s = n - 2 - 2t - 3q.$$

For each n value a degree partition matrix $M(n)$ is set up with $t^* + 1$ rows and $q^* + 1$ columns, where t^* and q^* are respectively the largest values of t and q such that the $(t + 1)$-th row and $(q + 1)$-th column of $M(n)$ each have a nonzero entry.

For computer programs which not only generate but also produce printouts of trees, see Section 5.12.

5.9 Polyhexes

Condensed polycyclic aromatic (benzenoid) hydrocarbons (polyhexes or PAH's) play a major role in chemistry, and their carcinogenicity has invested them with special interest. They were classified around 1880-1900 when their structures were elucidated into *cata*-condensed and *peri*-condensed polyhexes. The former (also called catafusenes) are considered to be devoid of carbon atoms common to three benzenoid rings; the latter (perifusenes) do have such carbon atoms. Balaban and Harary [55] proposed a new definition based on the dualist graph of such polyhexes. The dualist graph has as vertices the centres of benzenoid rings; the edges of the dualist graph connect vertices corresponding to condensed benzenoid rings, i.e. rings sharing a pair of adjacent carbon atoms. Unlike the dual graphs familiar to graph theorists, and unlike graphs in general, in dualist graphs bond angles are important. Polyhexes whose dualist graphs are acyclic are defined as catafusenes; those whose dualist graphs contain three-membered rings (possibly in addition to acyclic branches) are defined as perifusenes; those whose dualist graphs contain larger rings which are not perimeters of triangulated arrays (possibly in addition to three-membered rings and acyclic branches) are defined as coronafusenes (or after Dias [109] as circulenes). Examples are shown in the formulas 39–45. According to this definition, all catafusenes with the same number n of benzenoid rings are isomeric and have the formula $C_{4n+2}H_{2n+4}$.

Nonbranched catafusenes with n hexagons were enumerated by Balaban and Harary [52]; for odd n there are

$$(3^{n-2} + 4 \cdot 3^{n/2-3/2} + 1)/4 \qquad \text{isomers}$$

and for even n there are

$$(3^{n-2} + 2 \cdot 3^{n/2-1} + 1)/4 \qquad \text{isomers}.$$

For instance, for $n = 1, 2, 3, 4$, and 5 we obtain from the above formulas $1, 1, 2, 4$, and 10 isomeric catafusenes: benzene ($n = 1$), naphthalene ($n = 2$), phenanthrene and anthracene ($n = 3$), the four catafusenes, 39–42 with $n = 4$ shown in Figure 7, and the ten "pentacatafusenes" (i.e. with $n = 5$).

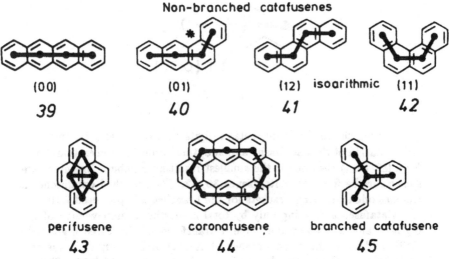

Non-branched catafusenes

| (00) | (01) | (12) isoarithmic | (11) |

39   40   41   42

perifusene   coronafusene   branched catafusene

43   44   45

Figure 7. Examples of polyhexes.

It should be stressed that all the above isomer numbers refer to a restricted class of isomers, namely those containing only six-membered (benzenoid) rings. If this restriction is relaxed to allow valence isomerism, benzene would lead to 6 cubic graphs on six vertices, and naphthalene would lead to 3838 valence isomeric graphs possessing eight vertices of degree three and two adjacent vertices of degree four. If the restriction were further relaxed to allow all possible isomers, benzene would lead to 217 graphs, as indicated in Section 5.8.

By definition, the dualist graph of a nonbranched catafusene has 2 vertices of degree one and $n - 2$ vertices of degree two; for branched catafusenes there exists at least one vertex of degree three.

A three-digit notation was developed [52,110–112] for coding the various catafusenes: an annelation at an angle of 180° is coded by the digit 0. (Figure 8), while annelations at angles of 120° or 240° are coded by the digits 1 or 2.

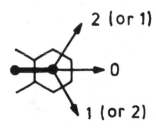

Figure 8.

Among the four possible sets of digits thus obtained (starting from either margin of the catafusene and adopting either the convention *1* or *2* for a left turn) the code is the smallest resulting number. Examples are shown for the four tetracatafusenes **39–42**. For branched catafusenes a similar code employing brackets for the branches was proposed [110].

Catafusenes differing only by total or partial interconversion of the digits *1* and *2* possess the same *L*-transform of the three digit code [113], and have the same number of Kekulé structures; such systems are therefore called isoarithmic (Greek: "same number") [114]. The *L*-transform replaces both *1* and *2* by *L*, and reads the resulting sequence of digits 0 and letters *L* as a binary number [113].

The number of zeros in the three digit code or in the *L*-transform is a significant parameter for catafusenes: it can be correlated [115] with resonance energies and is identical to the so-called number of imperfect benzenoid rings [116]. Thus in tetracene, i.e. (00) - tetracatafusene, 39, there are two benzenoid rings which cannot have three double bonds, in benzanthracene, i.e. (01) tetracatafusene, 40, there is one such ring, while in the two remaining isoarithmic catafusenes, 41 and 42 coded (11) and (12) respectively, all four benzenoid rings can be written with three double bonds. The number of zeros in the longest linear branch of a PAH can be correlated with the rate of Diels-Alder addition of maleic anhydride to the PAH [116].

The dualist graph code permitted a systematic computer search for PAH's possessing bay regions [117,118]. Such PAH's may display carcinogenic activity (a necessary but insufficient condition). A bay region in a PAH is defined as the region around a single benzenoid ring annelated with an angle of 120° or 240°. Examples are benzo-[a]-anthracene, 40, and benzo-[a]-pyrene, 46, where the bay region is indicated by an asterisk. Mitochondrial oxidation converts the isolated benzene ring adjacent to the bay region into an epoxydiol (diolepoxide 47) whose ring opening yields a stabilized benzylic cation, 48, which attacks the DNA strand leading to irreversible damage and then to malignant cell multiplication.

The formula for enumerating branched catafusenes is much more complicated [119]; so far no formula has been found for enumerating perifusenes, but computer programs have been developed for this purpose [120]. Trinajstić, Knop, Szymanski and coworkers [122] produced a program which not only generates isomeric polyhexes but also prints them out. A monograph containing 38202 structures with up to ten benzenoid rings was produced by Szymanski; some discrepancies between the latter results and previous ones [123] are due to the fact that the latter authors did not include coronafusenes or nonplanar helicenes. Perifusenes are much less tractable than catafusenes because they may not be isomeric even when they have the same number of benzenoid rings. Some perifusenes cannot have Kekulé structures; they are polyradicals, even if they have an even number of carbon atoms [112].

The enumeration of polyhexes has attracted several other chemists. Polansky and Rouvray [123] gave a formal graph-theoretical description leading to molecular formulas for polyhexes and to characterization of all-benzenoid PAHs in which each benzene ring has three double bonds. Dias [109a] developed further the grouping of polyhexes according to molecular formulas and produced a periodic table for PAHs. He also enumerated [109b] the 420 polycyclic conjugated isomers of pyrene having ring sizes ranging from three to nine.

The basic idea of dualist graphs and directions of annelation can assist in enumerating the isomers of polycyclic cata-condensed nonbenzenoid fully conjugated nonbranched systems [124]. Some of these have interesting molecular orbitals [127] despite the fact that the number of π-electrons is $4n + 2$, the alternant systems 49 and 50 have nonbonding MO's, whereas for nonalternants, system 51 has vacant bonding orbitals, system 52 has occupied antibonding orbitals, while system 53 has nonbonding orbitals. General rules were given for such systems [125]. Moreover, for such systems general formulas were found [124] giving the number of cata-condensed isomers as a function of ring size and order in which rings are annelated. For instance, there are two isomers with ring sizes 5,5,5,5 (one of which is 51), six isomers with ring sizes 7,7,7,7 (one of which is 52), and so on.

49

50

51

52

53

5.10 Diamond Hydrocarbons and Staggered Alkane Rotamers

The idea of the dualist graph based on the bidimensional graphite lattice can be extended to the tridimensional diamond lattice. In this case, *catamantanes* (e.g. adamantane, the diamantanes, etc.) have dualist graphs which are isomorphic to the staggered rotamers of alkanes. Such rotamer graphs may be coded by four digits (1,2,3,4) according to the four directions of the C–C bonds in the diamond lattice, and then adopting the convention of minimal resulting number [126–128]. Thus, staggered *n*-butane (which is the dualist graph of tetracatamantanes) can have either the *transoid* conformation, 54 (coded by 121), or one of the two enantiomeric *gauche* conformations, 55 (123) or 56 (124). Formulas were devised for enumerating all the possible staggered conformers of linear or branched alkanes, and of cycloalkanes [126].

(121)   (123)   (124)

54   55   56

The achiral tetracatamantane 57 has the dualist graph 54, while the chiral tetracatamantane 58 has the dualist graph 55 [127–128]. An interesting one-to-one correspondence was found between the three-digit codes of catafusenes and overlapping triplets from the four-digit codes of staggered alkanes rotamers. This implied equal numbers of isomers for nonbranched catafusenes with *n*-benzenoid rings and for linear alkane rotamers with $n + 1$ carbon atoms (with $n \leq 4$). For larger *n* values, differences appear because we allow rotamers, exemplified by 59, with "overlapping bonds" just as we allow nonplanar helicenes, exemplified by heptahelicene 60, since with slight bond distortions such molecules are able to exist. However, no such overlapping bonds are allowed in the dualist graphs of polymantanes. Therefore, starting with $n = 6$, we observe differences between the numbers of isomers in the staggered rotamer series and in the diamond hydrocarbon series.

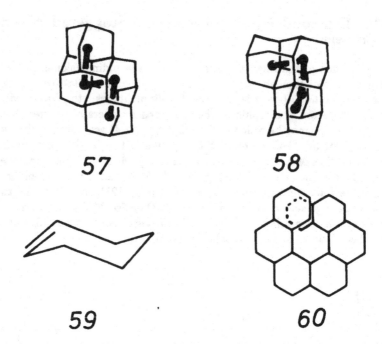

57   58

59   60

The results of calculations taking into account symmetry elements are presented in Table 9 [127].

Though all dualist graphs of diamond hydrocarbons are staggered alkane or cycloalkane C-rotamers, the converse is not true. Whenever in the four digit code of the dualist graph of the polymantane there is no sequence $axya$, (where a, b, x, and y are the set of digits 1, 2, 3, and 4 in any order), the polymantane will be regular and have formula $C_{4n+6}H_{4n+12}$. On the other hand, when such a sequence does exist, the polymantane has fewer carbon and hydrogen atoms than that required by the above formulas. When an acyclic C-rotamer has a sequence $abxab$ in its 4-digit code, no polymantane can correspond to such a dualist graph. Using a slightly different approach, Randić [129] was able to obtain confirmation of the numbers of alkane rotamers described above.

5.11 Diastereomeric Annulenes

Geometric isomers of the annulenes, i.e. of fully conjugated cyclopolyenes, are possible. An example is shown for [18]-annulene (only three

Table 9. Numbers of isomers for polyhexes, staggered alkane or cycloalkane rotamers, and polymantanes (diamond hydrocarbons)

| | Polyhexes | | | | | Staggered Rotamers | | | Diamond Hydrocarbons | | | | | | |
| | Catafusenes | | | Peri-fusenes | Poly-hexes | Alkanes | | Cyclo-alkanes | Total | Linear | | Branched | | Cyclic | Total |
n	Linear	Branched	Total			Linear	Branched			Reg	Irreg	Reg	Irreg	Irreg	
1	1	0	1	0	1	1	0	0	1	1	0	0	0	0	1
2	1	0	1	0	1	1	0	0	1	1	0	0	0	0	1
3	2	0	2	1*	3**	1	0	0	1	1	0	1	0	0	1
4	4	1	5	2*	7**	2	1	0	3	2	0	3	0	0	3
5	10	2	12	10*	22**	2	1	0	7	3	1	3	0	0	7
6	25	12	37	45*	82**	13	2	1	25	7	2	10	4	1	24
7	70	53	123	213*	333**	61	8	294	22	13	9	32	31	2	87

*These numbers count also radical or diradical systems, which have no Kekulé structures, and include non-planar heliceric systems. The numbers of non-radicalic perifusenes with $n = 3$ to 7 are: 0, 1, 7, 15, and 72, respectively.

**The numbers of non-radicalic polyhexes (with Kekulé structures) and $n = 3$ to 7 are: 2, 6, 19, 52, and 195, respectively.

diastereomers 61–63 are presented, all having bond angles of 120°) and for [16]-annulene (one configuration, **64**, having all bond angles of 120° but with five inner hydrogens; the other, **65**, having distorted bond angles but with lower steric repulsion because it has only four inner hydrogens).

61	*62A*	*62B*	*63*
(i) $(ctt)^3$	$(ctt)^3$	ccctcttct	ccctcttct
(ii) $(ctt)^3$	cctctccttcttcttctt		$(cctctcctt)^2$
(iii) $(121313232)^2$	1213132323131213232		$(121323213)^2$

64   *65*

The configuration of all double bonds may be coded by using the letters *c* for *cis* and *t* for *trans* [130]. This system, however, has two drawbacks: (a) the most serious is that it does not specify uniquely the configuration, as seen for **61** and **62A** or for **62B** and **63** (with **62A** and **62B** being limiting structures of the same system); (b) in some instances the two Kekulé structures of the same annulene configuration lead to different codes: this is the case for **62**, while **61** and **63** lead to the same code for both Kekulé structures in each case. One remedy is to specify the *c/t* configuration for each triad of bonds (overlapping triads), thereby doubling the number of letters in the code. A third possibility is to specify for each edge (only for configurations superimposable on the graphite lattice) one of the three possible directions in the plane by digits 1, 2 or 3 (which, as Gordon and Davidson [133] pointed out, are the only possible orientations of C-C bonds in such lattices). The last two coding systems remove the two drawbacks of the former code. In all three cases, it is assumed that *c t* and that 1 2 3, and one starts from the vertex and in the direction which leads to the smallest resulting code.

An enumeration of all possible configurations based on the third coding system was made [131].

5.12 Isomers and Computer Programs for Their Generation

The most general problem is to enumerate all possible isomers with a given molecular formula. From what was said above in the cases of restricted classes of isomers – which presented so many difficulties – it is clear that the general problem will be much more difficult. The only solution is to try to develop a computer program. This is exactly what Lederberg, Feigenbaum, Smith, Masinter, Sutherland and coworkers started to do in the 1960s [1,2,83,134,135].

They selected the problem of finding the structures of isomers as the most relevant and cost-efficient project in the area of artificial intelligence; an automated analysis of organic material by means of GC-coupled high-resolution mass spectrometry can readily afford the molecular formulas of the constituents. By finding all possible isomers, by selecting from these those structures which are chemically reasonable (see below), and by examining the remaining structures, using other methods such as NMR, it was anticipated that a universal identification method could be developed. Sponsorship of NASA was awarded since the project was presumably of interest for identifying organic materials of extraterrestrial origin. The result was an impressive HEURISTIC DENDRAL programme which met most of the requirements.

To exemplify the difficulties of devising a computer program for isomers, it is easy to see that an organic chemist given the molecular formula $C_6H_5NO_3$ will immediately write structures C_6H_5-O-NO_2, or the three *o*, *m*, *p*-HO-$C_6H_4NO_2$ isomers as the most plausible; a computer however, should be instructed that structures such as 66 are unstable; the chemist "knows" this is so, but the difficult problem was to prepare a BADLIST of unreasonable subgraphs.

Table 10 illustrates numbers of constitutional isomers calculated

$$HO-CH=C=CH-C\overset{C}{\underset{}{\diagup\diagdown}}C-O-O-NH_2$$

66

Table 10. Constitutional Isomers for given Molecular Formulas

Molecular Formula	Example compound	Number of Isomers
C_6H_6	Benzene	217
C_6H_8	Cyclohexadiene	159[a]
C_6H_{10}	Cyclohexene	77
C_6H_{12}	Cyclohexane	25[b]
C_6H_6O	Phenol	2237
$C_6H_{10}O$	Cyclohexanone	747
$C_6H_{12}O$	Hexanone	211
$C_3H_4N_2$	Pyrazole	155
$C_3H_6N_2$	Pyrazoline	136
$C_3H_8N_2$	Tetrahydropyrazole	62
$C_3H_{10}N_2$	Propylenediamine	14[b]

[a]Manual generation of isomers, tested by three chemists (one Ph.D. and two graduate students) did not give the correct result because of duplicates and omissions.
[b]Illustrated by Figure 9.

using the program of Masinter, Sridharan, Lederberg and Smith [135].
   The DENDRAL program is named so because it is based on a dendritic algorithm. Since multicyclic and polyfunctional molecules raise difficult problems, two daughter programs were devised. The CONGEN (constrained geometry) program combines several constitutional isomer generating functions, allowing the researcher to input structure constraints, thereby limiting the numbers and types of isomers to be generated. In this program, groups of atoms such as phenyl rings can be handled as

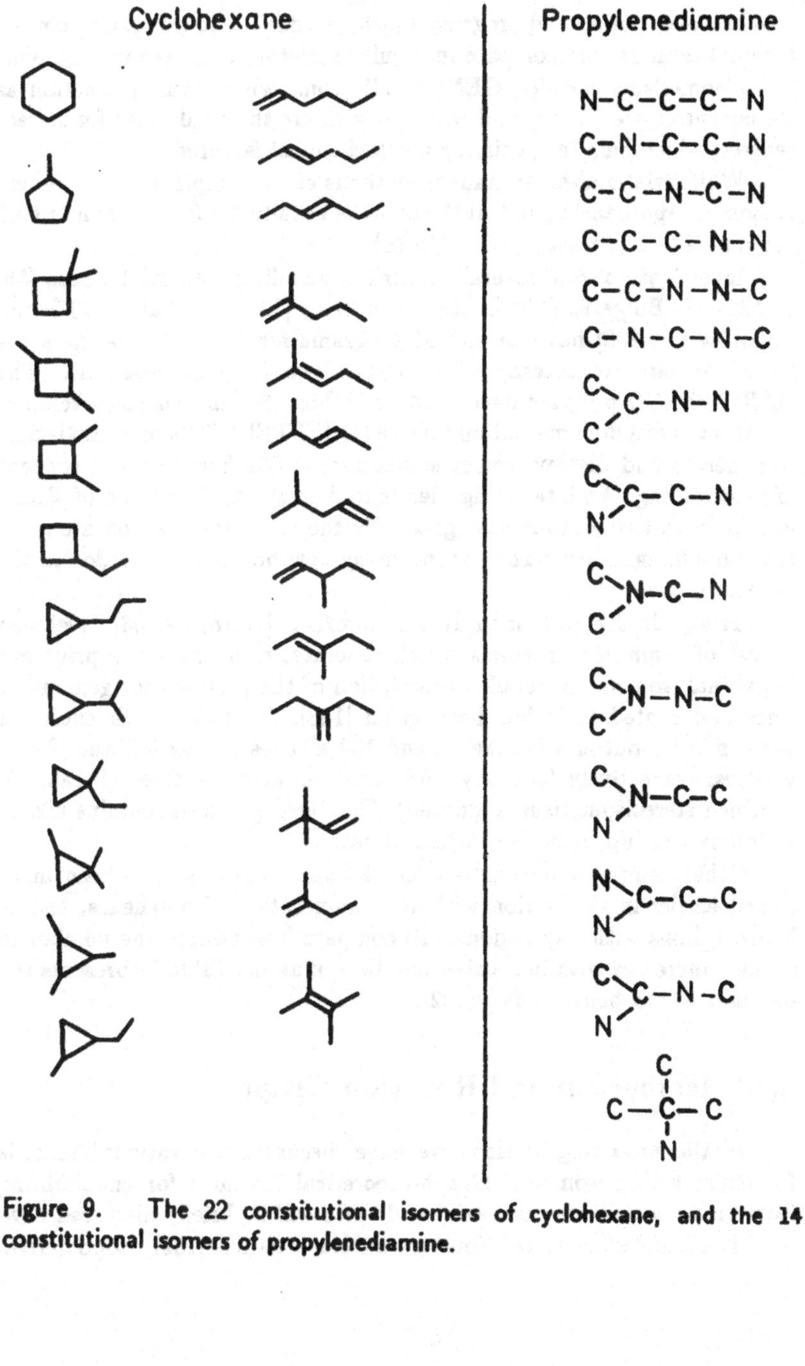

Cyclohexane

Propylenediamine

N–C–C–C– N

C–N–C–C– N

C–C –N–C–N

C–C– C– N–N

C–C–N –N–C

C– N–C– N–C

C\
  C–N–N\
C

C\
  C– C–N\
N

C\
  N–C– N\
C

C\
  N–N–C\
C

C\
  N–C–C\
N

N\
  C–C–C\
N

C\
  C–N –C\
N

    C\
C – C– C\
    N

Figure 9. The 22 constitutional isomers of cyclohexane, and the 14 constitutional isomers of propylenediamine.

"superatoms". A second program which, in addition to generating constitutional isomers, can compare molecular skeletons and take into account stereoisomerism, is called GENOA; all atoms which cannot function as stereocentres are left out and what remains are the candidates for stereocentres, which help in specifying stereochemical features.

While trial-and-error manual methods can be employed for numbers of isomers approaching 100, such methods usually fail for higher numbers of isomers (see footnote *a* of Table 10).

Investigators from several countries, including the Soviet Union [56, 84,136a,b], Bulgaria [136c], other countries [136d,e], France [137] and Yugoslavia [7,138] have developed programs for generating isomers. A few comments are necessary for the last two types of programs. The DARC-PELCO program developed by Dubois [5d] and his coworkers lies at the basis of a documentation service (QUESTEL). This program is able to generate and display isomer structures, and is based on the concept of focal coding. An interesting idea is to display the derivation of alkane isomers as a tree and increase gradually the number of carbon atoms so that on adding a new carbon atom, many new branches are added to the derivation tree.

Trinajstić, Knop, Symanski and coworkers [7,118,122,138] developed a host of computer programs which generate, enumerate and print out individual isomers. A detailed description of the program for generating trees and rooted trees has been given [138], illustrated with copies of the computer outputs for the 75 and 159 alkanes having ten and eleven vertices, respectively (actually these are the quartic 4-trees of Table 3, in which stereoisomerism is ignored). The book [10] also contains alkane skeletons with up to twelve carbon atoms.

Other computer programs from the same authors have been mentioned earlier in connection with the enumeration of polyhexes, and of hydrocarbons with bay regions. To compare how steeply the number of isomers increases with increasing numbers of atoms Table 11 presents the numbers of aza-benzenoids [7,122].

5.13 Isomerism and Reaction Graphs

In the preceding sections we have discussed the various methods for enumerating isomers. Graph-theoretical methods for enumerating isomerization reactions also exist and these will be briefly discussed now.

The simplest isomerizations are automerizations [139], (topomeriza-

Table 11. Numbers of isomeric aza-benzenoids [10,124].

	Number of nitrogen atoms								
n*	0	1	2	3	4	5	6	7	8
1	1	1	3	3	3	1	1	–	–
2	1	2	10	14	22	14	10	2	1
3	3	10	48	109	194	216	187	100	42
4	7	43	243	730	1620	2442	2802	2276	1410
5	22	210	1326	4918	12982	24611	35384	38500	32326
6	82	1026	7349	32043	98765	223717	388936	525764	561378

*Number hexagons

The original table involves up to ten hexagons. The printouts indicate structural
formulas of polyhexes, aza-polyhexes, etc. and the totals in the above table are
further subdivided according to the number of internal vertices, which are 0 for
catafusenes and > 0 for perifusenes.

tions [140]) in which the product is identical to the starting material,
though atom redistribution has taken place. Automerizations of pentasubstituted ethyl cations [141], like Berry pseudorotations of pentacoordinated phosphoranes [142], (trigonal bipyramids), can be depicted by the
same "reaction graph" in which the vertices are isomers, and the edges are
elementary steps. If one of the two carbon atoms of the pentasubstituted
ethyl cation can be distinguished by isotopic labelling, and if enantiomers
of pentasubstituted phosphoranes can be discriminated, there are 20 isomers in both cases if all five substituents are different [141,142], and the
reaction graph is a cubic non-planar graph (Desargues-Levi graph). If no
isotopic labelling for ethyl is present and if enantiomerism of phosphoranes
is ignored there are 10 isomers and the reaction graph is the nonplanar
cubic Petersen graph (5-cage), with a high symmetry [141,143].

Corresponding reaction graphs were devised for rearrangements of tetragonal pyramids [144], octahedral complexes [145] and automerizations
of various cations such as homocubyl [146]. Several analyses of the Berry
pseudorotation and of alternative mechanisms for the automerization of
phosphoranes have been reported [147-150]. Randić [151] investigated the
symmetry properties of such reaction graphs by means of his pathcode.
The whole field has been reviewed by Gielen [5d,152], and phosphorane
rearrangements by Luckenbach [153].

The most spectacular use of reaction graphs was in elucidating the mechanism of Schleyer's AlCl$_3$-catalyzed rearrangements of polycyclic compounds to the almost strain-free diamond hydrocarbons (adamantane, diamantane and higher homologues) [154]. Whitlock and Siefken [155] analyzed the mechanism of adamantane formation from the hydrogenated dimer of cyclopentadiene. Schleyer *et al* [156,157], and more recently Gander *et al* [158], analyzed the mechanism for the formation of diamantane. Interestingly, from Schleyer's analysis, in which the energetically most favourable reaction path at each step was computed, it was possible to infer the probable mechanism and to arrive at a plausible structure for a sideproduct which was isolated in amounts too small for a structure determination.

Balasubramanian reported a method for constructing isomerization reactions of molecules exhibiting large amplitude nonrigid motions [159]. The method not only enumerates the isomers of the nonrigid and the corresponding rigid molecules, but also the symmetry species of the isomers. By correlating the symmetry species of a group to the symmetry species of its subgroup, the splitting patterns of isomers of a nonrigid molecule to those of the corresponding rigid molecule are obtained; this is an elegant method for enumerating and constructing reaction graphs, e.g. for rotations of p-polyphenyls.

An interesting idea, due to Ugi and his coworkers, [5f,160] is to consider bond and electron matrices for sets of reacting molecules. Thus, any reaction, such as rearrangement, isomerization, addition, substitution or elimination, can be described in matrix terms. This idea allows matricial formulations (including computer storage and retrieval) of both chemical structures and reactions.

5.14 Conclusion

The first conclusion reached from a cursory glance through our preceding sections concerns the rapid increase of isomer numbers with increasing numbers of vertices of the skeleton graph, corresponding for instance to the number of carbon atoms in hydrocarbons. For alkanes, polyhexes and for cubic graphs the increase is slightly faster than exponential. The obvious inference is that one should have a rational nomenclature system for these rapidly proliferating isomers. Actually, while the nomenclature of acyclic systems is still manageable, for polycyclic systems (bridged or condensed) the present nomenclature is so cumbersome that trivial names continue to exist and even to proliferate (examples abound for the valence isomers of the annulenes). Suggestions for systematic coding and nomenclature have already been made [51, 52, 93, 96, 98, 101, 102], but a general system has yet to be devised and universally accepted.

Isomerism opened the gates for structure theory in organic chemistry, for Werner's theory of complexes in inorganic chemistry, and for stereochemistry. The above review has highlighted some aspects of isomer enumeration, with the inevitable bias of the author's interest in organic chemistry.

In the classical definition of isomers (Section 2.1), nothing was said about the stability of isomers, because the classical definition implies "stable at room temperature for a time long enough for isolation and measurement." This implies a half-life longer than seven hours at 300K (room temperature), and an interconversion barrier between isomers of at least 25 Kcal/mol [161]. However, NMR methods have time scales which can detect much shorter-lived species; with higher-energy electromagnetic radiation (ESR, microwave, IR, UV, X-rays, in increasing order) the barriers (i.e. the necessary stabilities) become lower and lower. The definition of isomers ought, therefore, to consider explicitly the stability in terms of temperature and time scale. There are two compelling reasons for revising correspondingly the definition of isomers. On the one hand, there is the advent of sophisticated spectrometric techniques (especially dynamic magnetic resonance), while matrix isolation at low temperatures permits the study of chemical species which were previously considered too unstable, such as rotamers. On the other hand, new structures are being synthesized which enhance interconversion barriers (to adopt the previous example, isolable rotamers at room temperature around single bonds have recently been synthesized). A new discussion is therefore opened in the area of isomerism, and implicitly in the field of isomer enumeration.

We have not discussed topological isomers such as rings versus knotted rings [162], or molecules without chemical bonds such as catenanes or rotaxanes [163,164]. Such species are known, and they give rise to interesting enumeration problems which have not yet caught the attention of graph-theorists.

Philosophically, isomer counting offers also an insight into the dimensionality of the microcosmos: down to the size of molecules, objects are tridimensional just like macroscopic objects. Indeed, an asymmetrically substituted tetrahedral molecule known to afford stable enantiomers would change its configuration if it could be "promoted" into the fourth dimension and returned to Euclidean tridimensional space. This parallels our showing that the two mirror image triangles 67 and 68 with different sides are congruent by lifting one of them and rotating it in tridimensional space before returning it to the plane of the paper. ·As far as the author is aware, no speculation on racemizations *via* "tetradimensional tunnelling" has yet appeared, and this may be due to the fact that even lower down the scale, namely in atomic nuclei, parity (i.e. chirality or handedness) is important. On the other hand, there have been many speculations on the possible tetradimensional character of a universe closed on itself like a hypersphere. Such speculations do not easily accommodate the big-bang theory, however.

congruent in 3D

67   **68**

and   congruent in 4D

In conclusion, isomer counting methods combine mathematical ingenuity with chemical insight and, owing to the advancement of computer hardware and software, are nowadays able to cope with almost any problem. However, since the combinatorics of organic chemistry afford an infinity of isomers (possibly of aleph one type) the challenge is still present to find more direct, fast and economic methods (in terms of manpower and computer time) for counting isomers.

References

1. J. Lederberg, Proc. Nat. Acad. Sci. USA, **53** (1965), 134.
2. J. Lederberg, in *The Mathematical Sciences*, MIT Press, Cambridge, Mass. 1969, p. 37.

3. D.H. Rouvray, Chem. Soc. Revs., **3** (1974), 355; Endeavour, **34** (1975), 28.

4. N. Trinajstić, *Chemical Graph Theory*, CRC Press, Boca Raton, Florida 1983.

5. A.T. Balaban (ed.), *Chemical Applications of Graph Theory*, Academic Press, London 1976; a) chapter 3 by F. Harary, E.M. Palmer, R.W. Robinson and R.C. Read; b) chapter 4 by R.C. Read; c) chapter 5 by A.T. Balaban; d) chapter 9 by M. Gielen; e) chapter 11 by J.E. Dubois; f) chapter 6 by J. Dugundji, P. Gillespie, D. Marquarding, I. Ugi and F. Ramirez.

6. Z. Slanina, *Teoreticke aspekty fenomenu chemickeisomeri*, Academia, Studie Ceskoslovenske Akademie Ved No. 4, Praha 1981.
   Z. Slanina, Adv. Quantum Chem., **13** (1981), 89.

7. J.V. Knop, W.R. Müller, K. Szymanski and N. Trinajstić, *Computer Generation of Certain Classes of Molecules*, Editions Kemija u industriji, Zagreb 1985.

8. E.L. Eliel, *Stereochemistry of Carbon Compounds*, McGraw-Hill, New York 1962; *Elements of Stereochemistry*, Wiley, New York 1969.

9. K. Mislow, *Introduction to Stereochemistry*, W.A. Benjamin, New York 1965.

10. H. Kagan, *Organic Stereochemistry*, Wiley, New York 1979 (French original, Press Universitaires de France, France, Paris 1975).

11. F. Badea and F. Kerek, *Stereochimie*, Editure Stüntifica, Bucharest 1974.

12. M. Nógradi, *Stereochemistry. Basic Concepts and Applications*, Akadémiai Kiado, Budapest 1981.

13. G. Natta and M. Farina, *Stereochemistry*, Longmans, London 1972 (Italian original, Ed. Sci. e Tecn., Mondadori 1968).

14. W. Bähr and H. Theobald *Organische Stereochemie: Begriffe und Definitionen*, Springer, Berlin 1973.

15. F.D. Gunstone, *Stereochemistry. A Programmed Course*, English Univ. Press, London 1974.

16. J. Pearce, *Stereochemistry: An Introductory Programme with Models*, Wiley-Interscience, New York 1978.

17. B. Testa, *Principles of Organic Stereochemistry*, Dekker, New York 1979.

18. V.M. Potapov, *Stereokhimiya*, Khimiya, Moscow 1976.

19. V.I. Sokolov, *Introduction to Theoretical Stereochemistry* (in Russian), Nauka, Moscow 1979.

20. J. Brocas, M. Gielen and R. Willem, *The Permutational Approach to Dynamic Stereochemistry*, McGraw-Hill, New York 1983.
21. IUPAC Joint Working Party on Stereochemical Definitions, "Stereochemical Terminology."
22. K. Mislow, Bull. Soc. Chim. Belges 86 (1977), 595.
23. F. Harary, *Graph Theory*, Addison-Wesley, Reading, Mass. 1979; R.J. Wilson, *Introduction to Graph Theory*, Longmans, London 1972; J.W. Essam and M.E. Fisher, Rev. Mod. Phys., 42 (1970) 272; D.H. Rouvray, R.I.C. Reviews 4 (1971), 173; Chem. Brit. 10 (1974), 11; J. Chem. Educ., 52 (1975), 768.
24. J.L. Gay-Lussac and L.J. Thénard, *Recherches Physico-Chimiques*, Deterville, Paris 1811, vol. 2, p. 340; J.L. Gay-Lussac, Ann. Chim., 91 (1814), 149.
25. J. Liebig, Ann. Chim. Phys., 24 (1823), 294; J.L. Gay-Lussac and J. Liebig, *ibid.* 25 (1825), 285.
26. M. Faraday, Phil. Trans. Roy. Soc., 115 (1825), 400.
27. J. Berzelius, Jahresbericht 1830; Poggendorfs Ann. 19 (1830), 326.
28. A.M. Butlerov, Z. Chem., 4 (1861), 549; 5 (1862), 298; 6 (1863), 500; 8 (1865), 614; Liebigs Ann. Chem., 144 (1867), 1; Bull. Soc. Chim. Fr. 5 (1866), 17.
29. A. Kekulé, *Lehrbuch der Organischen Chemie*, vol. 1, p. 183, Erlangen, (1861); Liebigs Ann. Chem., 106 (1858), 129.
30. A.S. Couper, Ann. Chim. Phys., 53 (1858), 469.
31. A. Crum Brown, Trans. Roy. Soc. Edinburgh, 23 (1864), 707.
32. J.A. LeBel, Bull. Soc. Chim. France, 22 (1874), 337.
33. J.H. Van't Hoff, Arch. Neer. Sci. Exact Nat., 9 (1874), 445.
34. A. Werner, Z. Anorg. Chem., 3 (1893), 267.
35. A. Cayley, Phil. Mag., (4), 18 (1859), 374; 13 (1857), 1, 172.
36. A. Cayley, Rep. Brit. Assoc. Adv. Sci., 45 (1875), 257; Ber. dtsch. chem. Ges., 183 (1875), 1056; Phil. Mag. (5) 3, 34 (1877); (4) 47 (1874), 444.
37. H. Schiff, Ber. dtsch. chem. Ges., 8 (1875), 1360, 1512.
38. F. Hermann, Ber. dtsch. chem. Ges., 13 (1880), 792; 30 (1897), 2423; 31 (1898), 91.
39. F. Tiemann, Ber. dtsch. chem. Ges., 26 (1893), 1595.
40. M. Delaunoy, Bull. Soc. Chim. France, 11 (1894), 239.
41. S.M. Losanitsch, Ber. dtsch. chem. Ges., 30 (1897), 1917, 3059.
42. H.R. Henze and C.M. Blair, J. Am. Chem. Soc., 53 (1931), 3042, 3077; 54 (1932), 1098, 1538; 55 (1933), 680; 56 (1934), 157;

D. Coffmann and C.M. Blair, *ibid.* 55 (1933), 252; D. Coffmann *ibid.* 55 (1932), 695.

43. D. Perry, J. Am. Chem. Soc., 54 (1932), 2918.
44. W. Burnside, *Theory of Groups of Finite Order*, 2nd Ed., p. 191, Cambridge Univ. Press, London 1911.
45. J.K. Redfield, Amer. J. Math. 49 (1927), 433.
46. A.C. Lunn and J.R. Senior, J. Phys. Chem., 33 (1929), 1027.
47. G. Pólya, Z. Kristal, (A) 93 (1936), 415; Acta Math. 68 (1937), 145; Helv. Chim. Acta, 19 (1936), 22.
48. E. Ruch, W. Hässelbarth and B. Richter, Theor. Chim. Acta, 19 (1970), 288; W. Hässelbarth and E. Ruch, *ibid.* 29 (1973), 259; A. Mead, E. Ruch and A. Schönhofer, *ibid.* 269; W. Hässelbarth and E. Ruch, Isr. J. Chem. 15 (1976/7), 112; W. Hässelbarth, E. Ruch, D.J. Klein and T.H. Seligman, Math. Chem. 7 (1979), 341.
49. N.G. De Bruijn, J. Combinatorial Theory, 2 (1967), 418.
50. F. Harary and E.M. Palmer, *Graphical Enumeration*, Academic Press, New York 1971; J. Combinatorial Theory, 1 (1966), 157.
51. A.T. Balaban, Rev. Roumaine Chim., 11 (1966), 1097; *erratum ibid.* 12, No. 1 (1967), last page.
52. A.T. Balaban and F. Harary, Tetrahedron, 24 (1868), 2505.
53. A.T. Balaban, D. Fărcasiu and F. Harary, J. Labelled Comp., 6 (1970), 211.
54. R.W. Robinson, F. Harary and A.T. Balaban, Tetrahedron, 32 (1976), 353.
55. A.T. Balaban and F. Harary, Rev. Roumaine Chim., 12 (1967), 1511.
56. M. Yu. Kornilov, Zhurn. Strukt. Khim., 8 (1967), 373.
57. C.C. Davis, K. Cross and M. Ebel, J. Chem. Educ., 48 (1971), 675.
58. A.T. Balaban, Studii Cercet. Chim. Acad. R.P. Romania, 7 (1959), 521.
59. E.K. Lloyd, in Proc. 1973 Brit. Combinat. Conf. at Aberystwyth (London Math. Soc. Lecture Notes Series).
60. J.G. Nourse, J. Am. Chem. Soc., 99 (1977), 2063; 101 (1979), 1210.
61. T.L. Hill, J. Phys. Chem., 47 (1943), 253, 413; J. Chem. Phys., 11 (1943), 294.
62. W.J. Taylor, J. Chem. Phys., 11 (1943), 532.
63. R.A. Kennedy, C.H. McQuarrie and C.M. Brubaker, Inorg. Chem., 3 (1964), 265.
64. T.E. Hass, Inorg. Chem., 3 (1964), 1053.

65. W.J. Klemperer, J. Am. Chem. Soc., 94 (1972), 6940, 8360; 95 (1973), 2105; Inorg. Chem., 11 (1972), 2668; J. Chem. Phys., 56 (1972), 5478.
66. G. Ege, Naturwiss., 58 (1971), 247.
67. W.S. DeLoach and J.B. Leroy, J. Elisha Mitchell Sci. Soc., 86 (1) (1970), 38; E.F. Welles and W.S. DeLoach, *ibid.*, 85 (2) (1969), 45; M.L. Shivar and W.S. DeLoach, *ibid.*, 84 (3) (1968), 368.
68. D.H. McDaniel, Inorg. Chem., 11 (1972), 2678.
69. R.A. Davidson and P.S. Skell, J. Am. Chem. Soc., 95 (1973), 6843.
70. R. Riemschneider, Z. Naturforsch., 11b (1956), 291.
71. W. Lehmann (A. Kerber, supervisor), Ph.D. Thesis, 1976.
72. R.C. Read, in *Graph Theory and Applications* (eds. Y. Ala, D.R. Lick and A.T. White), Springer, Berlin 1972, p. 243.
73. A.T. Balaban, Rev. Roumaine Chim., 20 (1975), 227.
74. D. Blackman, J. Chem. Educ., 50 (1973), 258.
75. W.H. Eberhardt, J. Chem. Educ., 50 (1973), 728.
76. R.L.C. Pilgrim, J. Chem. Educ., 51 (1974), 316.
77. A.T. Balaban, Croatica Chem. Acta, 51 (1978), 35.
78. J.E. Leonard, G.S. Hammond and H.E. Simmons, J. Am. Chem. Soc., 97 (1975), 5052.
79. R.B. Mallion, Croatica Chem. Acta, 56 (1983), 477.
80. J.R. Dias, Math. Chem., 13 (1982), 315.
81. A.T. Balaban, E.M. Palmer and F. Harary, Rev. Roumaine Chim., 22 (1977), 517; *erratum ibid.*, 23 (1978), 311.
82. A.T. Balaban and V. Baciu, Math. Chem. 4 (1978), 131.
83. L.M. Masinter, N.S. Sridharan, R.E. Carhart and D.R. Smith, J. Am. Chem. Soc., 96 (1974), 7714.
84. M. Yu. Kornilov, Zhur. Strukt. Khim., 16 (1975), 495.
85. R. Otter, Ann. Math., 49 (1948), 583.
86. K. Balasubramanian, Theor. Chim. Acta., 51 (1979), 37.
87. K. Balasubramanian, Theor. Chim. Acta., 59 (1981), 91.
88. R. Huisgen and F. Mietzsch, Angew. Chem. Internat. Edit. Engl. (1964).
89. A.T. Balaban, M. Banciu and V.T. Ciorba, *Annulenes, Benzo-, Hetero-, Homo-Derivatives and Their Valence Isomers*, CRC Press, Boca Raton 1985; G. Maier, *Valenzisomerisierungen*, Verlag Chemie, Weinheim 1972.
90. Woodward and Hoffman, Engl., 8 (1969), 781.
91. A.T. Balaban, Rev. Roumaine Chim., 18 (1973), 635; *erratum ibid.*, 19, No. 2 (1974), 338.

92. G. Maier, S. Pfriem, U. Schäfer and R. Matusch, Angew. Chem., 90 (1978), 552; G. Maier and S. Pfriem, *ibid.*, 90 (1978), 551.
93. A.T. Balaban, Rev. Roumaine Chim., 15 (1970), 463.
94. D.M. Walba, R.M. Richards and R.C. Haltiwanger, J. Am. Chem. Soc., 104 (1982), 3219; H.E. Simmons III and J.E. Maggio, Tetrahedron Lett., 22 (1981), 287; L.A. Paquette and M. Vazeux, *ibid.*, 22 (1981), 291; D.M. Walba, Tetrahedron, 41 (1985), 3161.
95. a) See e.g. the reverse face of the cover of J. Am. Chem. Soc. 95, Nos. 15, 16 (1973); b) A.H. Schmidt, Chem. unserer Zeit, 11 (1977), 118 and the cover of that issue.
96. A.T. Balaban and M. Banciu, J. Chem. Educ., 61 (1984), 766; M. Banciu, C. Popa and A.T. Balaban, Chem. Scripta 24 (1984), 28.
97. a) M. Avram, C.D. Nenitzescu and E. Marica, Chem. Ber., 90 (1957), 1857; b) P.E. Eaton, Y.S. Or and S.J. Branca, J. Am. Chem. Soc. 103 (1981), 2134; c) L.A. Paquette, Top. Curr. Chem., 79 (1979), 41; Proc. Nat. Acad. Sci. USA. 79 (1982), 4495; L.A. Paquette, R.J. Ternansky and D.W. Balogh, J. Am. Chem. Soc., 104 (1982), 4502; R.J. Ternansky, D.W. Balogh and L.A. Paquette, *ibid.*, 4503.
98. A.T. Balaban, Rev. Roumaine Chim., 17 (1972), 865.
99. A.T. Balaban, I. Motoc and R. Vancea, J. Chem. Inf. Comput. Sci., (to appear).
100. a) W. Imrich, Aequat. Math., 6 (1971), 6; b) R.E. Carhart, D.W. Smith, H. Brown and N.S. Sridharan, J. Chem. Inf. Comput. Sci., 15 (1975), 124.
101. A.T. Balaban, Rev. Roumaine Chim., 19 (1974), 1185.
102. M. Banciu and A.T. Balaban, Chem. Scripta, 22 (1983), 188.
103. V. Baciu and A.T. Balaban, Rev. Roumaine Chim., 23 (1980), 1213.
104. A.T. Balaban, Rev. Roumaine Chim., 19 (1974), 1185.
105. a) A. Ladenburg, Ber. dtsch. chem. Ges., 12 (1879), 141, 272; b) W. Körner and V. Wender, Gazz. Chim. Ital., 17 (1887), 5486.
106. A.T. Balaban, Rev. Roumaine Chim., 19, 1611; *erratum ibid.*, 23 (1978), 311.
107. A.T. Balaban, Rev. Roumaine Chim., 22 (1977), 987.
108. L.V. Quintás and J. Yarmish, Math. Chem., 12 (1981), 65; *ibid.*, 3 (1977), 2816; W.J. Martino, L.V. Quintas and J. Yarmish, *Degree partition matrices for rooted and unrooted 4-trees*, Pace University Report, Mathematics Department, New York 1980.
109. a) J.R. Dias, J. Chem. Inf. Comput. Sci., 22 (1982), 15, 139; b) J.R. Dias, Math. Chem., 14 (1983), 83.

110. A.T. Balaban, Tetrahedron 25 (1969), 2949.
111. A.T. Balaban, Math. Chem., 1 (1975), 33.
112. A.T. Balaban, Pure Appl. Chem., 54 (1982), 1075; Rev. Roumaine Chim. 26 (1981), 407, *erratum ibid.*, 27 (1982), 441.
113. A.T. Balaban, Rev. Roumaine Chim., 22 (1977), 45.
114. A.T. Balaban and I. Tomescu, Math. Chem., 14 (1983), 155; 12 (1981); Croat. Chim. Acta, 54 (1984), 391.
115. A.T. Balaban, Rev. Roumaine Chim., 15 (1970), 1243.
116. A.T. Balaban, D. Biermann and W. Schmidt, Nouv. J. Chim. 9 (1985), 443.
117. K. Balasubramanian, J.J. Kaufman, W.S. Koski and A.T. Balaban, J. Comput. Chem., 1 (1980), 149.
118. J.V. Knop, K. Szymanski, Z. Jericević and N. Trinajstić, Int. J. Quantum Chem. 23 (1983), 713.
119. F. Harary and R.C. Read, Proc. Edinburgh Math. Soc. (II) 17 (1980), 1.
120. D.A. Klarner, Canad. J. Math. 19 (1967), 851.
121. W.F. Lunnon, in *Graph Theory and Computing*, (ed. R.C. Read), Academic Press, New York 1972, p. 87.
122. N. Trinajstić, Z. Jericević, J.V. Knop, W.R. Müller and K. Szymanski, Pure Appl. Chem. 55 (1983), 379; K. Szymanski, "Polyhexagons der Ordnungen 1-10 klassifiziert nach der Anzahl der inneren Knoten," Univ. Düsseldorf Rechenzentrum (1982).
123. O.E. Polansky and D.H. Rouvray, Math. Chem., 2 (1976), 63, 91; 3 (1977), 97.
124. A.T. Balaban, Rev. Roumaine Chim., 15 (1970), 1251.
125. A.T. Balaban, Rev. Roumaine Chim., 17 (1972), 1531.
126. A.T. Balaban, Rev. Roumaine Chim., 21 (1976), 1049.
127. A.T. Balaban, Math. Chem., 2 (1978), 51.
128. A.T. Balaban and P. von R. Schleyer, Tetrahedron, 34 (1978), 3399.
129. M. Randić, Int. J. Quantum Chem. 7 (1980), 187.
130. J.F. M. Oth and J.M. Gilles, Tetrahedron Lett., (1968), 6259; J.F.M. Oth, G. Anthoine and J.M. Giles, *ibid.* (1968), 6265.
131. A.T. Balaban, Tetrahedron, 27 (1971), 6115.
132. A.T. Balaban, Rev. Roumaine Chim., 21 (1976), 1045.
133. M. Gordon and W.H.T. Davison, J. Chem. Phys., 20 (1952), 428.
134. J. Lederberg, G.L. Sutherland, B.G. Buchanan, E.A. Feigenbaum, A.V. Robertson, A.M. Duffield and C. Djerassi, J. Am. Chem. Soc., 91 (1969), 2973.

135. L.M. Masinter, N.S. Sridharan, J. Lederberg and D.H. Smith, J. Am. Chem. Soc., 96 (1974), 7702; B.G. Buchanan, A.M. Duffield and A.V. Robertson, in *Mass Spectrometry Techniques and Applications*, (ed. G.W.A. Milne), Wiley-Interscience, New York 1971, p. 121; H. Brown and L. Masinter, Discrete Math., 8 (1974), 227.

136. a) V.V. Raznikov and V.L. Talroze, Zhur. Strukt. Khim., 11 (1970), 357; b) V. Krivoshei, Zhur. Strukt. Khim., 8 (1967), 274; c) H. Dolhaine, Computers and Chem., 5 (1981), 41; d) W.E. Bennett, Inorg. Chem., 8 (1969), 1325.

137. J.E. Dubois, in *Computer Representation and Manipulation of Chemical Information* (eds. W.T. Wipke, S. Heller, R. Feldmann and E. Hyse), Wiley, New York 1974; J.E. Dubois and A. Panaye, Tetrahedron Lett. (1969), 1501, 3275; J.E. Dubois, Y. Sorel and C. Mercier, Comp. Rend. 292, II (1981), 783; F. de Closets, Chimie et Avenir No. 401, July 1980, p. 63, especially the diagram of alkanes with 1-12 carbon atoms on p. 69.

138. J.V. Knop, W.R. Müller, Z. Jericević and N. Trinajstić, J. Chem. Inf. Comput. Sci., 21 (1981), 91.

139. A.T. Balaban and D. Farcasiu, J. Am. Chem. Soc., 89 (1958), 967.

140. G. Binsch, E.L. Eliel and H. Kessler, Angew. Chem. Int. Ed. Engl., 10 (1971), 570.

141. A.T. Balaban, D. Farcasiu and R. Banica, Rev. Roumaine Chim., 11 (1966), 1205.

142. P.C. Lauterbur and F. Ramirez, J. Am. Chem. Soc., 90 (1968), 6722; M. Gielen and J. Nasielski, Bull. Soc. Chim. Belges, 78 (1969), 339; M. Gielen, C. Depasse-Delit and J. Nasielski, *ibid.*, 357; M. Gielen and C. Depasse-Delit, Theor. Chim. Acta, 14 (1969), 212.

143. J.D. Dunitz and V. Prelog, Angew. Chem. Int. Ed. Engl. 7 (1963), 725.

144. A.T. Balaban, Rev. Roumaine Chim., 23 (1978), 723.

145. A.T. Balaban, Rev. Roumaine Chim., 18 (1973), 841.

146. A.T. Balaban, Rev. Roumaine Chim., 22 (1977), 243.

147. J.I. Musher, J. Chem. Educ., 51 (1974), 94.

148. C. Zon and K. Mislow, Topics Curr. Chem., 19 (1971), 61.

149. E.K. Lloyd, Math. Chem., 7 (1979), 255.

150. A.T. Balaban, Rev. Roumaine Chim., 18 (1973), 855.

151. M. Randić, Chem. Phys. Lett. 42 (1976), 283; Int. J. Quantum Chem., 15 (1979), 663; *ibid.*, Quantum Chem. Symp. 14 (1980), 557; M. Randić and V. Katović, Int. J. Quantum Chem., 21 (1982), 647.

152. M. Gielen, *Stéréochimie dynamique*, Freund Publication House, Tel Aviv 1974.
153. R. Luckenbach, *Dynamic Stereochemistry of Phosphorus and Related Elements*, Thieme, Stuttgart 1973.
154. R.C. Fort, Jr., *Adamantane. The Chemistry of Diamond Molecules*, M. Dekker, New York 1976.
155. H.W. Whitlock and M. Siefken, J. Am. Chem. Soc., 90 (1968), 4929.
156. M. Gund, P. von R. Schleyer, P.H. Gund and W.T. Wipke, J. Am. Chem. Soc., 97 (1975), 743.
157. E.M. Engler, M. Farcasiu, A. Sevin, J.M. Cenze and P. von R. Schleyer, J. Am. Chem. Soc., 95 (1973), 5769; M. Farcasiu, E.W. Hagaman, E. Wenkert and P. von R. Schleyer, Tetrahedron Lett. 1501 (1981).
158. A.M. Kloster and C. Ganter, Helv. Chim. Acta, 66 (1983), 1200; F.J. Jäggi and C. Ganter, *ibid.*, 63 (1980), 866; A.M. Klester, F.J. Jäggi and C. Ganter, *ibid.*, 63 (1980), 1294.
159. K. Balasubramanian, Theor. Chim. Acta (in press).
160. I. Ugi, D. Marquarding, H. Klusacek, G. Gokel and P. Gillespie, Angew. Chem. Int. Ed. Engl., 9 (1970), 703; J. Gasteiger, P.D. Gillespie, D. Marquarding and I. Ugi, Topics Curr. Chem., 48 (1974), 1; I. Ugi, P. Gillespie and C. Gillespie, Trans. N.Y. Acad. Sci., 34 (1972), 415; J. Dugundji and I. Ugi, *ibid.*, 39 (1973), 19; I. Ugi, J. Dugurdji, R. Kopp and D. Marquarding, "Perspectives in Theoretical Stereochemistry," Lecture Notes in Chemistry No. 36, Springer, Berlin 1984.
161. V.I. Minkin, L.P. Olekhnovich and Yu A. Zhdanov, *Molecular Design of Tautomeric Systems* (in Russian), Izd. Rostovskogo Universiteta 1977.
162. H.L. Frisch and E. Wasserman, J. Am. Chem. Soc., 83 (1961), 3789; E. Wasserman, Sci. Am. 207 (1962), 94.
163. G. Schill, *Catenanes, Rotaxanes and Knots*, Academic Press, New York 1971.
164. I.S. Dimitriev, *Molecules Without Chemical Bonds* (in Russian) Khimia, Moscow 1980; *Moleküle ohne chemische Bindungen*, Verlag für Grundstaffindustrie Leipzig 1980.

Chapter 6

GRAPH THEORY AND MOLECULAR ORBITALS*

"They're two separate things, like wind and water;
they move each other, but don't mix."
Erich Maria Remarque: *Shadows in Paradise*

Nenad Trinajstić

The Rugjer Bošković Institute, P.O.B. 1016

41001 Zagreb, Croatia, Yugoslavia

---

*This article is dedicated to the memory of *Erich Hückel* (1896–1980)
and *Charles Alfred Coulson* (1910–1974), two great masters of theoretical
chemistry and mathematical chemistry.

6.1 Introduction

In the present article a graph-theoretical analysis of the simplest variant of molecular orbital theory [1], i.e. Hückel theory [2], will be given. Notwithstanding its naivety and shortcomings [3] Hückel theory is still extensively used in theoretical organic chemistry [4] in its various modifications [5]. This illustrates that quantum-chemical theories cannot be unconditionally judged as either good or bad, for much depends on the system under consideration, which of its physical, chemical, or biological properties are studied, and which approximation is used [6–10]. For these reasons, the theory which gives the greatest insight into a given chemical problem should be selected for use, bearing in mind that a qualitative rule covering a wide range of data is far better than a highly quantitative result which produces no general trends [11].

Because of its simplicity and limited computational endeavour, Hückel theory was especially useful in the pre-computer era of quantum chemistry (1931 – mid 1950's). During this time, first Lennard-Jones and Coulson, and later Coulson and Longuet-Higgins, in a series of papers, were able to mould Hückel theory into a mathematically consistent π-electron theory of unsaturated and aromatic molecules [5,12,13]. In the explosive computer era of quantum chemistry which has followed, many people have been misled into believing that heavy computation is the only way to treat chemical problems by theoretical methods. Consequently, two rather extreme views about the application of Hückel theory to chemistry have been promoted. One view is that the days of Hückel theory are numbered because there is no longer a need for such a naive and deficient theory, since high-speed computers and more sophisticated molecular orbital methods are now readily available [14]. The other view is that there is still room for Hückel theory in the chemistry of conjugated molecules on a qualitative level as a guide for chemists in the planning and interpretation of experiments [15]. This latter view is taken by most experimental chemists who prefer a "pencil and paper" method for their everyday research.

The last decade has produced a number of results which indicate that the successful longevity of Hückel theory is based on the fact that this theory contains intrinsic information about the internal connectivity in the conjugated structures, i.e. it reflects the neighbourhood of the atoms in the conjugated systems [5,16]. Here we will be concerned with the relationship between Hückel theory and the topology of the molecular π-network.

A mathematical tool for handling the topological problems of conjugated molecules is provided by *graph theory* [17], which is becoming a powerful method among the various mathematical methods available to theoretical chemistry. The topology of a molecular system is reflected essentially in the connections, i.e. chemical bonds, which exist between the atoms in the molecule. Hence, a *graph* is a convenient mathematical device for representing the topology of a given molecule because in the construction of the molecular (chemical) graph [16], the only property considered is the existence or non-existence of a chemical bond.

6.2 Elements of Graph Spectral Theory*

A graph, denoted by G, is a pair $(V(G), E(G))$, where $V(G)$ is a finite non-empty set of elements called *vertices*, and $E(G)$ is a finite set of unordered pairs of distinct elements of $V(G)$ called *edges*. A given graph may be depicted when its vertices are drawn as points and the edges are drawn as connecting lines. In a *molecular graph* the vertices represent atoms and the edges represent bonds of a molecule [16]. The molecular graph grossly simplifies the complex picture of a molecule by presenting only its *primary structure*, i.e. considering only the atom-atom connectivities and neglecting all other structural properties (e.g. geometry, symmetry, chirality, etc.). Conjugated molecules may be conveniently portrayed by a class of molecular graphs called *Hückel graphs* [16]. There is a one-to-one correspondence between the π-network of a given conjugated molecule and the corresponding Hückel graph. Hückel graphs are connected, undirected, and planar hydrogen-suppressed [18] graphs. The maximum valency of a vertex [17] in these graphs is 3. An example of a labelled Hückel graph, representing styrene, is given in Figure 1. A graph G is labelled if a precise numbering of the vertices in G is introduced.

A special type of Hückel graph is one which can be coloured in two

---

*This section overlaps, to a certain extent, the chapter "Polynomials in Graph Theory" by Dr. Ivan Gutman. Nevertheless we need to introduce in this section the definitions and concepts of graph spectral theory necessary for our further presentation. In Dr. Gutman's chapter all these are treated in great detail. The reader therefore is advised to consult his work for fine points on the theory of graph-theoretical polynomials and graph spectral theory, as they are only outlined briefly in this section.

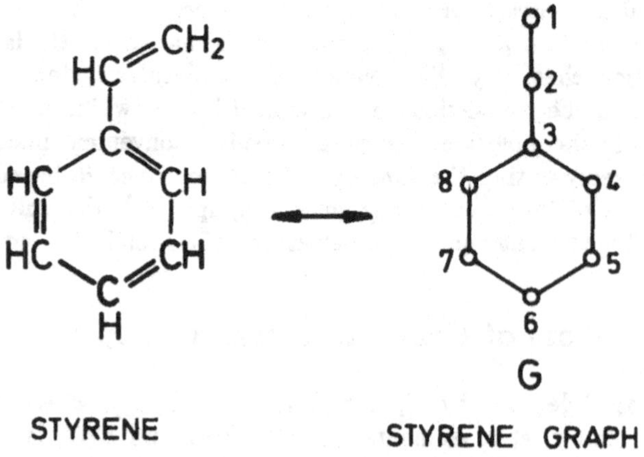

Figure 1. The Hückel graph of styrene

colours. Graphs which are bicolourable are called *bipartite graphs* or sometimes *bichromatic graphs*, whilst those that cannot be coloured in this way are called *non-bipartite graphs*. The colouring process is indicated by *stars* (*) and *circles* (o). Vertices of different "colours" in bipartite graphs are therefore called *starred* and *unstarred*. Let us denote the number of starred and unstarred vertices by s and u, respectively. By convention, $s \geq u$.

The graph-theoretical description of a bipartite graph G is as follows. Divide the set of vertices $V(G)$ into two non-empty subsets $V_1(G)$ and $V_2(G)$ in such a way that the first neighbours of the vertices in $V_1(G)$ are contained in $V_2(G)$ and vice versa, i.e. the vertices of the same colour are never adjacent if the graph is to be bipartite. The meaning of this is that if two vertices in the bipartite graph belong to the same (different) subset $V_i(G)$ ($i = 1, 2$), they are of the same (different) colour.

The actual colouring process is not needed for deciding whether or not a (Hückel) graph is bipartite, because of the following theorem due to König [19]: *A graph is bipartite if, and only if, all its cycles (rings) are even-membered.*

Conjugated hydrocarbons that may be depicted by Hückel bipartite graphs are called *alternant hydrocarbons*, AH's, whilst *non-alternant hydrocarbons*, NAH's, are represented by non-bipartite graphs.

Alternant structures may be classified as *acyclic* and *cyclic*. Each of these classes may be partitioned into *even* and *odd*. Even alternant structures have even number of atoms, whereas the odd alternant structures have odd numbers of atoms. The even cyclic alternants may be further divided into two groups depending on whether $N/2$ (N = number of atoms) is even or odd.

The $N/2$ = odd cyclic alternants consist of a combination of $4m + 2$ ($m = 1, 2, \ldots$) rings, whilst the $N/2$ = even cyclic alternants represent a combination of $4m$ and $4m + 2$ rings, respectively. If we consider only monocyclic alternants, these fall into two groups: $4m$ and $4m + 2$. In chemistry these structures correspond to [N]annulenes [20], where N = even. [N]annulenes with N = odd also fall into two classes: [$4m + 1$]- and [$4m + 3$]-annulenes.

We note in Figure 1 that all *adjacent vertices* are connected. In general, two vertices in G are adjacent if they are joined by an edge. This may be suitably stated in matrix form. A given graph, adequately labelled, may be associated with the *adjacency matrix* [16,21]. The (vertex) adjacency matrix $\mathbf{A}(G) \equiv \mathbf{A}$, is a real symmetric $N \times N$ matrix defined as,

$$(\mathbf{A})_{ij} = \begin{cases} 1 & \text{if vertices i and j are adjacent} \\ 0 & \text{otherwise.} \end{cases} \tag{1}$$

The graph G is uniquely characterized by its adjacency matrix, but the reverse is not, in general, true because the labelling of the vertices of G is arbitrary [22].

The adjacency matrix of, for example, the labelled styrene graph in Figure 1 is given by,

$$\mathbf{A}(G) = \begin{bmatrix} 0 & 1 & 0 & 0 & 0 & 0 & 0 & 0 \\ 1 & 0 & 1 & 0 & 0 & 0 & 0 & 0 \\ 0 & 1 & 0 & 1 & 0 & 0 & 0 & 1 \\ 0 & 0 & 1 & 0 & 1 & 0 & 0 & 0 \\ 0 & 0 & 0 & 1 & 0 & 1 & 0 & 0 \\ 0 & 0 & 0 & 0 & 1 & 0 & 1 & 0 \\ 0 & 0 & 0 & 0 & 0 & 1 & 0 & 0 \\ 0 & 0 & 1 & 0 & 0 & 0 & 1 & 0 \end{bmatrix} \tag{2}$$

If a bipartite graph is suitably labelled, the corresponding adjacency matrix appears in block-form [23]. Let us label a bipartite graph G in such a way that vertices $1, 2, \ldots, p$ belong to subset $V_1(G)$, whilst vertices

$p+1, p+2, \ldots, p+q(=N)$ are in the subset $V_2(G)$. Then $(A)_{su} = 0$ for $1 \leq s$, $u \leq p$ and $p+1 \leq s$, $u \leq p+q$, because the vertices of the same colour are never adjacent in the bipartite graph. Hence the adjacency matrix of a bipartite graph adopts the block-form,

$$
A(G) = \begin{bmatrix} 0 & B \\ B^T & 0 \end{bmatrix}
\tag{3}
$$

where B is a submatrix of $A(G)$ with dimensions $p \times q$ and B^T is its transpose.

The adjacency matrix of a graph could be submitted to various transformations [24]. The linear transformation leading to the *diagonal form*, X, is of the greatest importance,

$$
X = \begin{bmatrix}
x_1 & & & & \\
& x_2 & & & O \\
& & \ddots & & \\
& & & \ddots & \\
O & & & & \ddots \\
& & & & x_N
\end{bmatrix}
\tag{4}
$$

Elements $x_1, x_2, \ldots, x_N$ of a diagonal matrix X are the eigenvalues of A. By convention,

$$
x_1 \geq x_2 \geq \cdots \geq x_N.
\tag{5}
$$

If some eigenvalue appears k times, it is a k-fold degenerate. The set of all eigenvalues $x_1, x_2, \ldots, x_N$ is collectively called the *spectrum* of a graph. The graph spectrum is an important graph invariant [22]. Since A is a symmetric matrix by definition, a linear transformation can always be carried out, and furthermore, the elements of the graph spectrum are all real numbers.

The transformation to diagonal form is realized by the *eigenvector matrix* C of A

$$
C\,A = X\,C.
\tag{6}
$$

The eigenvector matrix C is an $N \times N$ matrix of the form,

$$
C = \begin{bmatrix} C_1 \\ C_2 \\ \vdots \\ C_N \end{bmatrix}
\tag{7}
$$

where C_i $(i = 1, 2, \ldots, N)$ denotes a row matrix of C,

$$C_i = [c_{i1}, c_{i2}, \ldots, c_{iN}]. \tag{8}$$

Eq. (7) may be rewritten as,

$$C_i A = x_i C_i; \qquad i = 1, 2, \ldots, N \tag{9}$$

where C_i are the eigenvectors belonging to the eigenvalue x_i. This equation may be rewritten in the form,

$$C_i(x_i I - A) = 0; \qquad i = 1, 2, \ldots, N \tag{10}$$

where I is the unit matrix. Eq. (10) represents a system of homogeneous linear equations, that is, equations whose constant terms are zero. They are also called the *secular equations* and may be expressed, as below, in a more explicit format,

$$\sum_{k=1}^{N} c_{ik}[x_{ik}\delta_{kl} - (A)_{kl}] = 0; \qquad i, l = 1, 2, \ldots, N \tag{11}$$

where δ_{kl} is the Kronecker delta function. In order that the secular equation (11) have non-trivial solutions, it is necessary that the corresponding *secular determinant* vanishes,

$$\det |x_i I - A| = 0; \qquad i = 1, 2, \ldots, N. \tag{12}$$

A polynomial,

$$P(G; x) = \det |x_i I - A| = 0; \qquad i = 1, 2, \ldots, N \tag{13}$$

is called the *characteristic polynomial* (of the adjacency matrix) of a graph. $P(G; x)$ is the polynomial of degree N,

$$P(G; x) = \sum_{n=0}^{N} a_n(G)x^{N-n} \tag{14}$$

where $a_n(G) \equiv a_n$ $(n = 0, 1, \ldots, N)$ are the coefficients of the polynomial. The set of all zeros of the characteristic polynomial makes up the graph spectrum.

The characteristic polynomial may be constructed, and the graph spectrum obtained, in several alternative ways.

Here we will briefly describe the procedure based on *the coefficients theorem of Sachs* or, in short, *Sachs' theorem* [25]. This is a very general, elegant, and imaginative way of constructing the coefficients of the characteristic polynomial. However, the method is not really practical for large graphs because with the size of the system there is an explosion of combinatorial possibilities which even large computers cannot handle [26]. Nevertheless, it reveals directly the structural basis of the characteristic polynomial, it is used for deriving the related polynomials [27], and it forms the basis for the study of isospectral (molecular) graphs [28]. Sachs method has been selected here because later in this article it will be used for deriving the *acyclic polynomial*, a basic quantity in the topological resonance energy model [29].

Before formulating Sachs' theorem, let us define *the Sachs graphs* [30] (Grundfigure, basic figure, mutation graph, characteristic graph). We call a subgraph s of G a Sachs graph if every component (elementary figure) [25] of it is either a complete graph K_2 or a cycle C_m $(m = 3, 4, \ldots, N)$ or a combination of $1 \cdot K_2$ and/or $r \cdot C_m$, with the restriction $2 \cdot 1 + r \cdot m = n$. ($K_2$ is a *complete graph* of valence one. A graph is a complete graph of valence one if it consists of just two vertices joined by an edge). Let $c(s)$ and $r(s)$ denote, respectively, the total number of components and the total number of cycles in s. Now we can state the Sachs theorem: *The coefficients of a characteristics polynomial of a given graph G can be obtained from the following formula,*

$$a_n = \sum_{s \in S_n} (-1)^{c(s)} 2^{r(s)}; \qquad 0 \le n \le N \qquad (15)$$

where S_n is the set of all Sachs graphs with n vertices of G. The summation in (15) is over all elements of the set S_n. For the case n = 0, $a_0 = 1$ (by definition). If S_n is an empty set, $a_n = 0$.

Sachs' theorem can be extended to *vertex-* and *edge-weighted graphs*. Heteroconjugated molecules may be represented by vertex- and edge-weighted graphs [16], denoted by G_{VEW}. A vertex- and edge-weighted graph G_{VEW} is a graph which has one or more of its vertices and edges distinguished in some way from other vertices and edges. These vertices and edges of different "type" are weighted and their weights are identified by parameters h (weighted vertices) and k (weighted edges) for heteroatoms and heterobonds, respectively. An example of the vertex- and edge-weighted hydrogen suppressed Hückel graph is given in Figure 2.

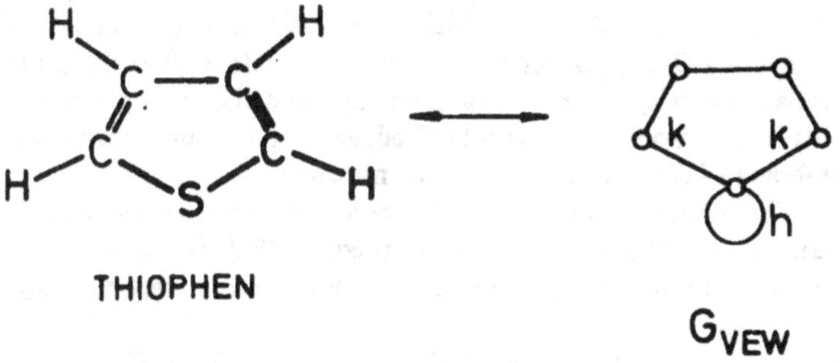

Figure 2. The vertex- and edge-weighted hydrogen-suppressed Hückel graph of thiophene.

The use of parameters h and k is reflected on some diagonal and off-diagonal elements in the adjacency matrix of G_{VEW}. The adjacency matrix corresponding to G_{VEW} in Figure 2 is given below:

$$\mathbf{A}(G_{VEW}) = \begin{bmatrix} h & k & 0 & 0 & k \\ k & 0 & 1 & 0 & 0 \\ 0 & 1 & 0 & 1 & 0 \\ 0 & 0 & 1 & 0 & 1 \\ k & 0 & 0 & 1 & 0 \end{bmatrix} \qquad (16)$$

Before stating the coefficients theorem for vertex- and edge-weighted graphs we need to introduce a new type of Sachs graph containing weighted vertices and/or weighted edges. It is also immediately evident that a graph G_{VEW} may contain both types of Sachs graphs, that is to say, non-weighted and weighted graphs. We define *a vertex-and/or edge-weighted Sachs graph* as such a subgraph of G_{VEW} which has no components other than one-cycles (loops) and/or (weighted) K_2 graphs and/or (weighted) cycles C_m. We can now state the coefficients theorem for the vertex- and edge-weighted graphs: *The coefficients of a characteristic polynomial of a given weighted graph G_{VEW} can be obtained from the following formula:*

$$a_n = \sum_{s \in S_n} (-1)^{c(s)} 2^{r(s)} \prod_i^s h_i \prod_j^{K_2 \text{ in } s} k_j^2 \prod_{j'}^{C \text{ in } s} k_{j'}; \qquad 0 \le n \le N. \quad (17)$$

In the above equation, symbols have the following meaning: h_i is the weight of the i-th vertex depicted by loop, the first product gives the contribution from all the weighted vertices i with weights h_i in s, k_j is the weight of the j-th edge, which may be either a K_2-component of s or an edge in a C-component of s, the second product gives the contribution from all the weighted K_2-components in s and the third product gives the contribution from all the weighted edges in the C-component of s. Other symbols in (17) have their previous meaning.

The interval over which the zeros of $P(G; x)$ are distributed is bounded. According to Frobenius' theorem [31], the limits of the graph spectrum are defined by the maximum valency of a vertex, D_{max}, in G,

$$-D_{max} \leq x_i \leq D_{max}; \qquad i = 1, 2, \ldots, N. \tag{18}$$

The meaning of (18) is that the quantities x_i may or may not reach their extreme values, but they never fall outside the interval $(-D_{max}, +D_{max})$. As a rule the x_i's are not integers, in contrast to D_{max}, but some graphs have spectra of whole numbers [32].

6.3   The Essence of Hückel Theory

The theoretical framework of Hückel theory has often been presented in the literature [3,5,8,12,13]. Therefore, we will give here only its bare outline.

Hückel theory was devised to treat electrons in unsaturated molecules. In Hückel theory, only the π electrons are considered explicitly. This is the result of the Hückel approximation of $\sigma - \pi$ separability,

$$\int \psi_\sigma \cdot \psi_p \, d\tau = 0 \tag{19}$$

where ψ_σ and ψ_π represent the σ and π electronic systems of a conjugated molecule, respectively.

The Hückel approximation is based in the following qualitative argument. A conjugated molecule is in general a planar, or near-planar, system. In this very special situation, molecular orbitals, MO's, describing the molecule may be partitioned into two orthogonal groups: σ MO's which are symmetric and π MO's which are antisymmetric to reflection in the plane of the molecule. Physically, the Hückel approximation may be viewed as one which has the π electrons moving in a potential field due to the nuclei and a σ core, which is assumed to be rigid as the π electrons move around. The Hückel approximation, when introduced, had a three-fold justification. Firstly, with this bold approximation, Hückel reduced an insurmountable computational problem to one which was feasible in those pre-computer days. For example, benzene, instead of the formidable problem of 30 valence electrons converts into a problem of only 6 electrons. Secondly, Hückel found that, by treating only the π electrons explicitly, it is possible to reproduce many of the observed properties of unsaturated molecules [2,33]. Finally, organic chemists have related quite correctly the physical and chemical properties of conjugated molecules (thermodynamic and structural parameters, spectroscopic features, reactivity) to the presence of their π electrons. In addiiton, the subsequent work by a large number of researchers [3,5,8,12,13] has also supported the Hückel approximation, because it has revealed many useful correlations between the quantities obtained from Hückel theory and experimental findings.

The Hückel molecular orbitals, HMO's, of a conjugated molecule are eigenfunctions ψ_i of the effective one-electron Hamiltonian, called the Hückel Hamiltonian, $\hat{H}$(Hückel), which is defined only by its matrix elements [5],

$$\hat{H}(\text{Hückel})\psi_i = E_i\psi_i; \qquad i = 1,2,\ldots,N \tag{20}$$

where E_i is the energy eigenvalue associated with ψ_i. N stands for the number of atoms in a molecule bearing π electrons.

The individual Hückel orbital ψ_i is expressed as a linear combination of atomic orbitals, LCAO [34],

$$\psi_i = \sum_{r=1}^{N} c_{ir}\phi_r \tag{21}$$

where c_{ir} are the linear expansion coefficients, whilst ϕ_r is a $2p_z$ orbital on atom r. The summation is over all atoms r in a conjugated molecule.

The total π electron energy, E_π, is given by

$$E_\pi = \sum_{i=1}^{N} g_i E_i \tag{22}$$

or more explicitly,

$$E_\pi = \sum_{i=1}^{N} g_i \frac{\sum_r \sum_s c_{ir}^* H_{rs} c_{is}}{\sum \sum c_{ir}^* c_{is} S_{rs}} \tag{23}$$

where g_i is the occupation number of ψ_i, i.e. the number of electrons that populate ψ_i, whilst H_{rs} and S_{rs} are short-hand notations for the integrals,

$$H_{rs} = \int \phi_r^* \widehat{H}(\text{Hückel}) \phi_s d\tau \qquad (24)$$

$$S_{rs} = \int \phi_r^* \phi_s d\tau. \qquad (25)$$

The coefficients c_{ir} are obtained from the requirement that E_π should be a minimum. Minimization of E_π by means of the variational procedure leads to a set of simultaneous, linear, homogeneous equations,

$$\sum_{t=1}^{N} c_t (H_{rt} - E_i S_{rt}) = 0; \qquad i, r = 1, 2, \ldots, N. \qquad (26)$$

If this set of secular equations is to have non-trivial solutions (the trivial solution has all $c_{it} = 0$), the corresponding Hückel (secular) determinant must vanish,

$$\det |H_{rt} - E_i S_{rt}| = 0; \qquad i, r = 1, 2, \ldots, N. \qquad (27)$$

The Hückel determinant can be simplified by using the set of approximations originally introduced by Bloch [35] and utilized by Hückel [2]. Since the Hückel Hamiltonian is not known explicitly, the backbone of the *Bloch-Hückel approximations* is the presumption that the entries to the Hückel determinant may either be related to empirical quantities or removed entirely.

The Bloch-Hückel approximations are discussed below.

(i) *The diagonal elements, H_{rr}*

$$H_{rr} = \int \phi_r^* H(\text{Hückel}) \phi_r d\tau = \alpha. \qquad (28)$$

The diagonal elements H_{rr} are *Coulomb integrals* with empirical values α. It is assumed that the α's are constant for all orbitals ϕ_r centered on similar atoms r, regardless of the variations in the neighbouring atoms and groups.

(ii) *The off-diagonal elements, H_{rs}*

The off-diagonal elements H_{rs} are assumed to be zero unless orbitals ϕ_r and ϕ_s are located on bonded atoms,

$$H_{rs} = \int \phi_r^* \hat{H}(\text{Hückel}) \phi_s d\tau = \begin{cases} \beta & \text{if atoms r and s are bonded} \\ 0 & \text{otherwise.} \end{cases} \tag{29}$$

For bonded atoms the off-diagonal elements H_{rt} are called *resonance integrals* with empirical value β. It is assumed that the β's have the same value for π bonds between the same kind of atoms, regardless of the environment.

(iii) *The overlap integral, S_{rs}*

$$S_{rs} = \int \phi_r^* \phi_s d\tau = \begin{cases} 1 & \text{if } r = s \\ 0 & \text{if } r \neq s. \end{cases} \tag{30}$$

This is the so-called *zero-overlap approximation* which is rather drastic because it says that there is no overlap between the atoms making up the π network of a conjugated molecule. However, the neglect of overlap is justified empirically by the success of Hückel theory over the past fifty years. The inclusion of overlap among the bonded atoms changes the spacing of energy levels and the values of the total π electronic energies, but other quantities remain unchanged [36].

The parameters α and β for carbon atoms (α_C and β_{CC}) are taken to be the reference points. The introduction of a heteroatom X into the conjugated system alters both of these parameters. These changes are expressed as,

$$\alpha_X = \alpha_C + h_X \beta_{CC} \tag{31}$$

$$\beta_{CX} = k_{CX} \beta_{CC} \tag{32}$$

where h_X and k_{CX} are dimensionless parameters for a given heteroatom in a specific molecular environment. The ways of selecting the parameters h_X and k_{CX} are presented in many publications [37] and are discussed even nowadays [38].

6.4   Isomorphism of Hückel Theory and Graph Spectral Theory

Eq. (27) may be presented in a more compact, and elegant, form if matrix notation is used. Thus, Eq. (27) in matrix form is given by,

$$\det |H - E_i S| = 0; \qquad i = 1, 2, \ldots, N \tag{33}$$

where H and S are the Hamiltonian and overlap matrices, respectively. As a result of the Bloch-Hückel approximations, the matrices H and S have the following composition [39],

$$H = \alpha I + \beta A \tag{34}$$

$$S = I \tag{35}$$

where A is the adjacency matrix of the Hückel graph (conjugated molecule).

Substitution of H and S by (34) and (35) into (33) and, dividing each row of a determinant by β, leads to,

$$\det \left| \frac{E_i - \alpha}{\beta} I - A \right| = 0; \qquad i = 1, 2, \ldots, N. \tag{36}$$

If the normalized form of Hückel theory is used, i.e. if β is taken as the energy unit and α the zero-energy reference point, $\beta = 1$ and $\alpha = 0$, equation (36) becomes,

$$\det |E_i I - A| = 0; \qquad i = 1, 2, \ldots, N. \tag{37}$$

Comparison between the secular determinant (12) and the Hückel determinant (37) reveals that the numbers E_i, representing the energies of individual Hückel orbitals, are *identical* to the elements of the spectrum of eigenvalues of a given Hückel graph,

$$E_i = x_i; \qquad i = 1, 2, \ldots, N. \tag{38}$$

Since matrices H and A commute (this can easily be seen from (34)),

$$H A - A H = 0 \tag{39}$$

they have the same eigenvectors. Therefore, the eigenvectors of the adjacency matrix are *identical* to the Hückel MO's. On account of this the Hückel molecular orbitals are sometimes called the *topological orbitals*. Expression (34) also discloses that the Hückel Hamiltonian is a function of the adjacency matrix [40],

$$H = H(A). \qquad (40)$$

This is due to the singular nature of the Hückel Hamiltonian, with the short-range forces being dominant in the effective potential [41].

The above analysis leads to two important conclusions:

(1) The spacing and general pattern of Hückel eigenvalues are specified by the connectivity ("topology") in the conjugated molecule, and

(2) The connectivity ("topology") in the conjugated molecule, rather than its geometry, shapes up the form of Hückel orbitals.

Thus, what chemists commonly refer to as Hückel theory is essentially the same thing as graph spectral theory (in the area of planar connected undirected graphs with maximum valency 3) referred to by graph theoreticians. In fact Hückel theory and graph spectral theory are isomorphic theories [5,16,22,40,42].

At this point we remind the reader that in order to derive various quantities from the knowledge of Hückel orbitals and energies (charge density, bond order, free valence, etc.), we also need to know the electronic configuration appropriate to the ground state (or to any other state, if required) of the molecule under consideration [43]. In order to assign the appropriate electronic configuration, the fundamental physical notion of the *Aufbau principle* [44] must be called upon. According to the Aufbau principle, each of the energy levels in a molecule can be populated by a maximum of two electrons of opposite spin in accordance with the *Pauli exclusion principle*. The electrons are fed in from the lowest to the highest energy level until all of the available electrons are used up. In case when degeneracies occur in the energy levels the filling up process also involves the use of *Hund's rules*.

6.5 The Spectrum of a Hückel Graph

The Hückel spectrum is given by an ordered sequence of numbers, the extreme values of which are defined by Frobenius' theorem [31],

$$x_1, x_2, \ldots, x_N. \qquad (41)$$

Since the maximum valency in Hückel graphs is 3, the interval which limits all the elements in the Hückel spectrum is,

$$-3 \leq z_i \leq +3; \qquad i = 1, 2, \ldots, N. \qquad (42)$$

Because the maximum graph theoretical valency in the linear polyenes and annulenes is 2, the extreme values in the Hückel spectra for these systems are ± 2.

The Hückel spectrum may be separated into three subsets corresponding to the bonding, N_+, non-bonding, N_0, and anti-bonding, N_-, MO levels. They are related to the total number of atoms N in the conjugated molecule,

$$N_+ + N_0 + N_- = N. \qquad (43)$$

These quantities are important in the chemistry of conjugated molecules. For example, the presence of non-bonding molecular orbitals, NBMO's, in the Hückel spectrum of a given molecule is indicative of its low stability and high reactivity [45]. It is indeed an established fact that the structures possessing NBMO's rarely occur in the chemistry of conjugated hydrocarbons [46], and even then these are obtained under the drastic conditions of low temperature chemistry [47].

A Hückel spectrum may be used for the classification of conjugated hydrocarbons [48]. The relationship between the numbers N_+, N_0, and N_- leads to the prediction of *four* classes of conjugated hydrocarbons:

(i) *Normal molecules* distinguished by,

$$N_+ = N_- \qquad (44)$$

$$N_0 = 0. \qquad (45)$$

Molecules in this class are expected to be stable. However, the above criterion is *necessary*, but not sufficient for a given molecule to be stable in a chemical sense. The prediction of the stability of a chemical compound is an extremely difficult problem involving thermodynamic and kinetic aspects. Therefore, the above may serve only as a suggested indicator for neutral molecules whose stabilities may vary from very stable molecules to rather unstable species.

(ii) *Polyradical molecules* distinguished by,

$$N_+ = N_- \qquad (46)$$

$$N_0 \geq 0. \qquad (47)$$

Structures in this class should be very reactive.

(iii) *Electron deficient molecules* distinguished by,

$$N_+ > N_- \tag{48}$$

$$N_o \geq 0. \tag{49}$$

Structures belonging to this class should exhibit a tendency to generate stable anions by accepting π electrons from suitable donors in their empty MO bonding levels.

(iv) *Electron-excessive molecules* distinguished by,

$$N_+ < N_- \tag{50}$$

$$N_o \geq 0. \tag{51}$$

Structures in this class should show a tendency to produce stable cations by releasing π electrons from the anti-bonding MO levels.

This classification is in fair agreement with experimental findings [49–55]. The above is, of course, a very simple-minded approach, and it should be used cautiously.

6.6 The Number of Non-Bonding Molecular Orbitals

It is important to establish the presence of non-bonding molecular orbitals, NBMO's in conjugated structures because their existence leads to the prediction [45] that such molecules should have open-shell ground states and be very reactive. Although in reality the situation is much more complicated (for example, because of Jahn-Teller effects in the case of triplet ground states), it is an established fact that structures possessing NBMO's are rarely encountered in the chemistry of conjugated hydrocarbons [46].

In the mathematical and chemical literature there are a number of studies of the occurrence of NBMO's [16,45,56]. These approaches are considerably diverse in their application, some of them being particularly impractical for larger molecules, whilst others are limited only to one very narrow class of conjugated molecules. Here we will outline a simple method for the determination of NBMO's. Since we know that the number of NBMO's is identical to the number of zeros, N_o, in the graph spectrum, and since,

$$\det \mathbf{A} = \sum_{i=1}^{N} x_i. \tag{52}$$

The determinant of $\mathbf{A}$ will be zero if, and only if, there exists at least one zero in the spectrum. Therefore, the problem of determining N_0 is closely related to the problem of evaluating the determinant of the adjacency matrix of the graph in question. There are several recipes available in the literature [57-59] for the evaluation of the determinant of the adjacency matrix of a molecular graph. In the present work we will discuss two procedures. The first is purely a numerical procedure and consists of the diagonalization of the adjacency matrix and the inspection of the spectrum. The other procedure is graph-theoretical [58]. Before stating the final result we have at this point to introduce and discuss several novel concepts.

Dewar and Longuet-Higgins [57] have found an important relationship between Hückel theory and resonance theory: *The determinant of* $\mathbf{A}$ *can be evaluated from a knowledge of the number of Kekulé structures of a conjugated molecule whose graph has the adjacency matrix* $\mathbf{A}$. Kekulé structures are principal resonance structures in the qualitative valence bond theory. The concept of Kekulé structures coincides with the concept of 1-factors in the mathematical literature [17]. The number of Kekulé structures is denoted by K. Dewar and Longuet-Higgins found that in the case of alternant hydrocarbons, Kekulé structures may be separated into two classes of different parity: *even*, K^+, and *odd*, K^-, respectively. The same (even) or different (odd) parity is determined in the Dewar-Longuet-Higgins scheme according to whether the number of transpositions of double bonds required to transform one of the structures into the other is *even* or *odd*. There is some difficulty in determining the parity of Kekulé structures by use of the Dewar-Longuet-Higgins method [60]. Accordingly, we have endeavoured to establish simple rules for determining the mutual parity of two Kekulé structures [16,58,60,61]. For an arbitrary conjugated system (subject to the limitation that it should not contain only odd-membered rings if it is a three- or many-cyclic system) the following rule can be used for determining the parity of two Kekulé structures: *If the Sachs graph obtained by superposition of two Kekulé graphs contains an even (odd) number of 4m-membered rings, the Kekulé structures in question have the same (opposite) parity.* Kekulé graphs depict Kekulé structures [62]. The Kekulé structure count is defined as,

$$K = K^+ + K^- \tag{53}$$

whilst the *algebraic structure count* [63] (corrected structure count [64]) ASC, is given by

$$ASC = K^+ - K^-, \qquad K^+ > K^-. \tag{54}$$

K^+ and K^- are related to the determinant of the adjacency matrix in the following way [58],

$$\det A = (-1)^{N/2}(K^+ - K^-)^2 + (-1)^N \sum_{s \in S_N^*} (-1)^{c(s)} 2^{r(s)} \tag{55}$$

where S_N^* is the set of Sachs graphs which are simultaneously spanning subgraphs of a graph and which contain at least one odd-membered cycle. (*A spanning subgraph* of G is a subgraph containing all the vertices of G [17].) For alternant hydrocarbons which do not contain odd-membered rings [19], $S_N^* = \emptyset$ the above equation reduces to the Dewar-Longuet-Higgins formula [57],

$$\det A = (-1)^{N/2}(K^+ - K^-). \tag{56}$$

The application of formulae (55) and (56) will be illustrated in Figure 3.

Hence, the problem as to whether a conjugated system has or has not NBMO's can be solved; another problem, however, is to enumerate the NBMO's. Again the numerical approach consisting of the diagonalization of the adjacency matrix of G is the simplest. However, by using this we do not learn much about the (structural) origin of NBMO's. Therefore, we will delineate the Longuet-Higgins-Živković algorithm [45,56] for enumeration of N_o (NBMO's).

If $C = (c_1, c_2, \dots, c_N)$ is an NBMO (not necessarily normalized), the following equation holds,

$$C\,A = 0. \tag{57}$$

This equation may also be given in scalar form,

$$\sum_{i>j} c_i = 0; \qquad j = 1, 2, \dots, N \tag{58}$$

where the summation is over all vertices i joined to the vertex j. The above equation represents a *zero-sum rule* first used by Longuet-Higgins [45]. The number of independent parameters in an unnormalized

Figure 3. The application of (55) to s-indacene
(1) Kekulé structures of s-indacene

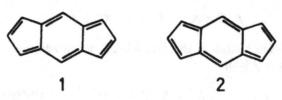

1   2

(2) Kekulé graphs

K_1 K_2

(3) Determination of parity
    Structure 1 is arbitrarily selected to be even (+1)

$$K_1 = +1$$

$K_1 \cdot K_2 =$ ⬡ ⟶ $K_2 = -1$

Supperposition graph
(4m-cycle)

$$K^+ = 1, \qquad K^- = 1.$$

Figure 3. The application of (55) to s-indacene (continued)

(4) Kekulé structure count and algebraic structure count

$$K = K^+ + K^- = 2$$
$$\text{ASC} = K^+ - K^- = 0$$

(5) Determination of S_N^*
Since K_1 and K_2 are only possible Sachs spanning subgraphs of s-indacene and since they do not contain odd-membered cycles,

$$S_N^* = \emptyset.$$

(6) $\det A = 0$.
(7) Prediction
s-indacene must have at least one NBMO.

NBMO is equal to the number N_0 in the Hückel spectrum [56]. Thus, the enumeration of N_0 is reduced to a determination of the number of independent parameters in NBMO which satisfy the zero-sum rule. The application of this method is illustrated for s-indacene (Figure 4). The procedure is as follows: eq. (58) is stepwise satisfied for each vertex of a graph G. Vertices for which the zero-sum rule is executed are denoted by black dots (•).

6.7 Total Pi-Electron Energy

The total π-electron energy, E_π, is one of the most important parameters of a conjugated molecule that can be obtained from the Hückel MO calculations [65]. It can be used, selectively, to relate to the thermodynamic stability of conjugated structures. A parametrization scheme, based on thermodynamic data, by Hess and Schaad [66], has produced agreement with experiment to the same degree of quantitative accuracy as the much more sophisticated SCF π-MO procedure developed by Dewar [67]. In addition, Schaad and Hess [68] have shown that in many instances E_π follows linearly the total (thermodynamically measurable) energy of the conjugated molecule. The physical reasons leading to the answer as to how it is possible that a model as simple as the Hückel model can give not only qualitative, but sometimes also fair quantitative

Figure 4.   The enumeration of NBMO's in s-indacene

(1) Stepwise application of the zero-sum rule (58)

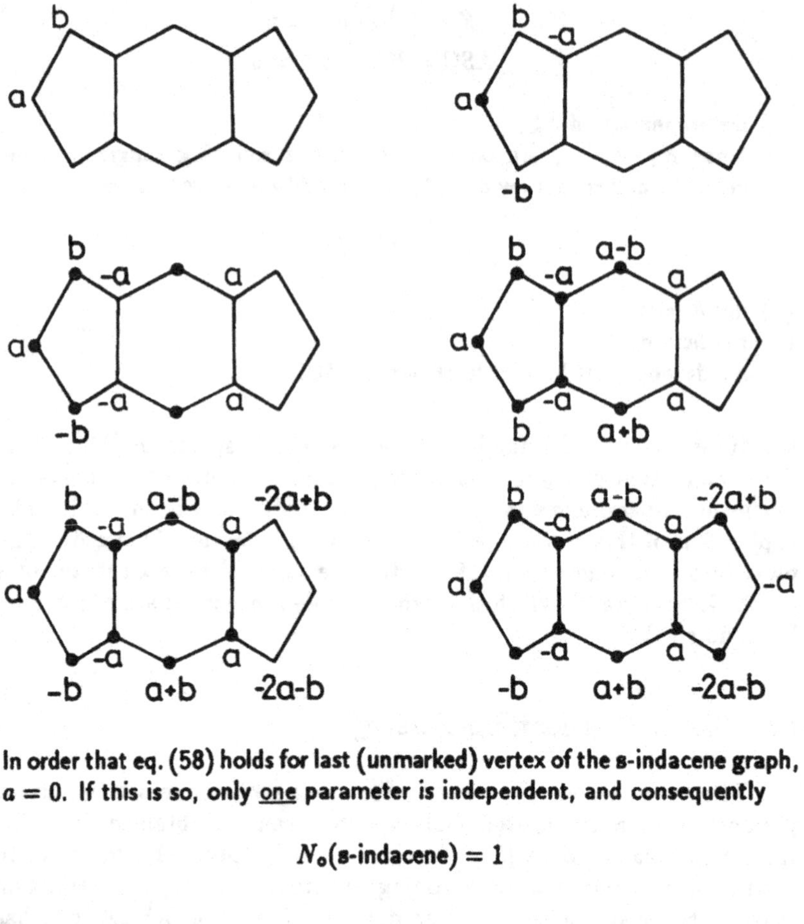

In order that eq. (58) holds for last (unmarked) vertex of the s-indacene graph, $a = 0$. If this is so, only <u>one</u> parameter is independent, and consequently

$$N_o(\text{s-indacene}) = 1$$

agreement with experimental findings, are not well understood at present.

The total π-electron energy for a conjugated molecule with N π-electrons is given by,

$$E_\pi = \sum_{i=1}^{N} g_i x_i \tag{59}$$

Figure 4. The enumeration of NBMO's in s-indacene (continued)

(2) The normalized NBMO of s-indacene

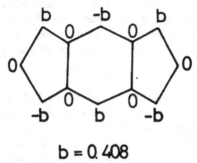

$$b = 0.408$$

This result is in agreement with traditional Hückel calculations [33].

where g_j is the number of electrons (0, 1, or 2) in the j-th MO.

Let the considered conjugated molecule with N atoms and N_π electrons be in the ground state with a closed-shell configuration. Then the total π-electron energy is defined as follows,

$$E_\pi = \begin{cases} 2\sum_{i=1}^{N_\pi/2} x_i, & \text{if } N_\pi = \text{even} \\ 2\left[\sum_{i=1}^{(N_\pi-1)/2} x_i + x_{(N_\pi+1)/2}\right], & \text{if } N_\pi = \text{odd}. \end{cases} \tag{60}$$

However, in the great majority of conjugated systems $N = N_\pi$ and therefore, the above formulae become,

$$E_\pi = \begin{cases} 2\sum_{i=1}^{N/2} x_i & \text{if } N = \text{even} \\ 2\sum_{i=1}^{(N-1)/2} x_i + x_{(N+1)/2} & \text{if } N = \text{odd}. \end{cases} \tag{61}$$

Several researchers [69–73] have attempted to derive a topological formula which would enable the estimation of the total π-electron energy when some details of molecular topology, expressed in terms of graph invariants [17] (e.g. number of atoms, number of bonds, number of 4-, 6-, ..., n-membered rings, etc.), are known. However, Gutman [74] has recently shown that *no* exact expression for E_π can be derived in terms of graph invariants (topological parameters) only.

The alternative approach is to find an approximate relationship between E_π and the topological parameters. Here we will discuss several recent achievements in this area. The first work that we wish to mention is by McClelland [75]. He has produced the following relationship,

$$E = a(2MN)^{1/2} \tag{62}$$

where N and M are the number of atoms and the number of bonds, respectively, in a conjugated molecule, whilst $a = 0.92$ is an empirical constant. McClelland's result reveals that the gross part of E_π is determined by M and N. Numerical calculations show that M and N make up not less than 95% of E_π. All other topological factors play, therefore, a seemingly marginal role. However, it is to be noted that in all chemical applications we are interested not solely in energies but energy differences. The problem of the remaining 4-5% of E_π is thus essential for chemistry. Thus, McClelland's formula is not very useful for discussions about, for example, the ordering of isomers, because all isomers have the same value of McClelland's E_π [76]. Nevertheless, McClelland's formula is useful in discussions concerning the dependence of E_π on the size of molecules. For example, it can be shown that it is not correct to compare the E_π values of molecules with the same N, but different number of rings, R. The number of rings in a molecule is given by,

$$R = M - N + 1. \tag{63}$$

Therefore, eq. (62) may be rewritten as,

$$E = a[2N(R + N - 1)]^{1/2}. \tag{64}$$

This means that E_π is also proportional to the number of rings present in the molecule.

McClelland's formula was later improved by Gutman [77,78] who left out the empirical constant but included the algebraic structure count,

$$E_\pi = (2MN)^{1/2} - \frac{1}{2}(AB^3)^{1/4} \tag{65}$$

where

$$A = (N/2) \cdot \ln(2M/N) - 2 \cdot \ln(\text{ASC}) \tag{66}$$

$$B = \frac{1}{2} \sum_{i=1}^{N} d_i^2 + 2n_4 - M^2/N - M/2 \tag{67}$$

where n_4 is the 4-membered ring whilst d_i is the degree of the vertex i, i.e. the number of its first neighbours. In Hückel graphs $d_i = 1, 2,$ or 3. Gutman's formula exhibits a dependence of E_π on the number of atoms, the number of bonds, the algebraic structure count, branching and the number of 4-membered rings in the molecule. Branching, as a structural characteristics of a given molecule enters, to (65) through the valencies of the vertices. The greater number of branched vertices [79], that is vertices with $d_i = 3$, the larger is $D = \sum_{i=1}^{N} d_i^2$. Therefore, D may be understood as a measure of the branching of the π-electron network.

Another approximate non-empirical formula which connects E_π and the number of atoms, the number of bonds, and the algebraic structure count has also been derived by Gutman [78],

$$E_\pi = N + [2(M - N/2)\ln(\mathrm{ASC})]^{1/2}. \qquad (68)$$

Several formulae of high accuracy which relate E_π and N, M and ASC, by least-squares have been obtained by the Zagreb Group [16,80,81].

Starting from the Dewar-Longuet-Higgins result [57], expressed as (56), a three-parameter topological formula for the total Hückel energy was obtained [81],

$$E_\pi = AN + BM + C\ln(\mathrm{ASC}) \qquad (69)$$

with the values of the coefficients $A = 0.913$, $B = 0.347$, and $C = 0.765$ obtained by a least-squares fitting to an arbitrarily selected group of conjugated hydrocarbons.

The three-parameter formula for E_π (69) confirms the conclusion already reached earlier by McClelland, and later by Gutman, that E_π is primarily determined by N and M. However, the term $\ln(\mathrm{ASC})$ is responsible for subtle effects, such as conjugative stabilization (or destabilization). Namely, for a set of structural isomers, having the same number of atoms, bonds, the rings of the same size, and the same type of skeletal branching, E_π increases linearly with $\ln(\mathrm{ASC})$. Hence, it follows that ASC's may be used as a sufficient criterion for predicting relative stabilities (since E_π is related to the thermodynamic stability of a conjugated system [68] of structurally closely related isomers). This result also explains why traditional resonance theory was so successful in the thirties. In those days only benzenoid hydrocarbons were known. (The question concerning cyclooctatetraene has not been yet settled). Due to the fortunate fact that, for benzenoids, $K = $ ASC, resonance theory predictions based on K were in all cases correct. The trouble started when the chemistry of non-benzenoid hydrocarbons developed. Then, a correction of the resonance theory was needed and this was provided by the ASC concept.

Note when ASC $= 0$ or $K = 0$, relations (65), (68), and (69) need corrections to account for the appearance of NBMO's.

This work has disclosed that besides N and M, the greatest contribution to E_π comes from the (algebraic) structure count. Therefore, we can safely state a simple rule for predicting the stability order of the set of n structural isomers as,

$$\text{ASC}(1) \geq \text{ASC}(2) \geq \cdots \geq \text{ASC}(n) \implies E(1) \geq E(2) \geq \cdots \geq E(n).$$
$$(70)$$

The above result was known to Dewar and Longuet-Higgins [57], and has been discussed, though not so explicitly, by several authors [81,82]. Exceptions of the above rule are found only in the case of large branched polyenes [83].

A different topological formula has been proposed for E_π by Hosoya *et al* [84]. A modified topological index of Hosoya [85] defined as,

$$\tilde{Z}(G) = \sum_{n=0}^{[N/2]} (-1)^n a_{2n} \tag{71}$$

is proposed for characterizing conjugated hydrocarbons. In eq. (71) a_{2n} is the coefficient of the characteristic polynomial of the Hückel graph G. Calculation of $\tilde{Z}(G)$ is exemplified in Figure 5.

Hosoya and co-workers [84] found that E_π values are linearly related to the logarithms of $\tilde{Z}(G)$,

$$E_\pi = A \log \tilde{Z}(G) + B \tag{72}$$

where the least-squares constants A and B are for (a) linear polyenes: 6.092 and 0.129, (b) annulenes: 6.092 and 0.000, and (c) polyacenes: 6.041 and 0.242, respectively.

Aihara [86] found that there is a linear relationship between E_π and $Z^*(G)$ of the form,

$$E_\pi = A \log Z^*(G) \tag{73}$$

Figure 5.   Calculation of the $\widetilde{Z}(G)$ index for benzene and fulvene

$$[N/2] = 3$$

G

$$P(G;x) = x^6 - 6x^4 + 9x^2 - 4$$

$$\widetilde{Z}(G) = (-1)^0 a_0 + (-1)^1 a_2 + (-1)^2 a_4 + (-1)^3 a_6$$

$$= 1 + 6 + 9 + 4 = 20$$

$$[N/2] = 3$$

G

$$P(G;x) = x^6 - 6x^4 + 8x^2 - 2x - 1$$

$$\widetilde{Z}(G) = (-1)^0 a_0 + (-1)^1 a_2 + (-1)^2 a_4 + (-1)^3 a_6$$

$$= 1 + 6 + 8 + 1 = 16$$

where $Z^*(G)$ is a topological index close to the modified Hosoya index $\widetilde{Z}(G)$, whilst $A(6.0846)$ is an empirical constant. Aihara's $Z^*(G)$ index is defined as,

$$Z^*(G) = \prod_{i=1}^{N}(1 + x_i^2)^{1/2} \qquad (74)$$

where x_i's are zeros of the characteristic polynomial of G. For any alternant hydrocarbon, $Z^*(G)$ is an integer and agrees with $\widetilde{Z}(G)$. However, for any non-alternant hydrocarbon, $Z^*(G)$ is never integer and hence disagrees with $\tilde{Z}(G)$.

6.8 Topological Resonance Energy

The resonance energy, RE, is a rather enduring concept [12] which has served for many years as a theoretical measure of the aromaticity of conjugated systems [87,88]. The RE concept has been the raison d'être for the continuous use of Hückel theory by organic chemists. For example, many recent developments of Hückel theory were indeed stimulated by a desire to remedy the failure of traditional forms of the theory in predicting the aromatic behaviour of conjugated molecules.

The RE may be understood as the difference between the total π electron energy of a conjugated molecule and the π electron energy of some reference structure,

$$RE = E_\pi(\text{molecule}) - E_\pi(\text{reference structure}). \tag{75}$$

The main difficulty with the RE concept is the hypothetical nature of the reference structure: its choice being somewhat arbitrary. The RE index of conjugated hydrocarbons, calculated in the standard way,

$$RE = E_\pi - 2n_{C=C} \tag{76}$$

has not always been a reliable criterion for aromaticity, because very unstable molecules such as pentalene or heptalene are predicted to be aromatic [29,66]. $n_{C=C}$ in eq. (76) denotes the number of double bonds in one Kekulé structure of the molecule under study. The failure of the classical formula for RE computations (76) may be understood from the following argument. Let us introduce $n_{C=C} = N/2$ in (76),

$$RE = E_\pi - N. \tag{77}$$

Then, by utilizing McClelland's formula (64), eq. (77) becomes

$$RE = a[2N(R + N - 1)]^{1/2} - N. \tag{78}$$

Assuming that $N \gg R$, eq. (78) becomes,

$$RE = N(a\sqrt{2} - 1) = 0.3N. \tag{79}$$

This relationship reveals that the classical RE index is grossly proportional to N and thus insensitive to other details of molecular structure which may be directly related to the aromatic behaviour of a given molecule.

Dewar [67] was the first to amend eq. (76) satisfactorily, though there have been other, less successful, attempts in this direction [89]. Dewar introduced the *polyene reference structure* instead of the usual isolated double reference structure. This new RE index was termed the *Dewar resonance energy* [90], DRE,

$$DRE = E_\pi - (n_{C-C}E_{C-C} + n_{C=C}E_{C=C}) \qquad (80)$$

where n_{C-C} and $n_{C=C}$ are the number of single and double bonds, respectively, in a Kekulé structure of the conjugated molecule, whilst E_{C-C} and $E_{C=C}$ are parameters usually interpreted as the π electron energies of the polyene single and double bonds. Dewar's model, therefore, is based on the acyclic polyene-like reference structure. Dewar and coworkers, using this model have calculated DRE indices of many hundreds of conjugated molecule and have obtained, in the majority of cases, good predictions of aromaticity [91]. These calculations were carried out within the framework of an original variant of the SCF π-MO method [67]. However, the decisive step was the change of the reference structure and not, in fact, the use of the more advanced MO method [92]. The application of Dewar's concept to Hückel theory has produced excellent agreement with experimental findings [4,66,68,92,93].

Dewar's definition of RE is based on the possibility of approximating the π electron energy of polyenes by summing up all polyene bond energies,

$$E(\text{polyene}) = \sum_{i=1}^{k} n_i E_i \qquad (81)$$

where n_i and E_i are the number of energies, respectively, of a particular bond type appearing in the polyene.

Milun *et al* [94] reported the calculation of aromatic stabilities, A_s, using the standard form of Hückel theory, in which the acyclic polyene-like reference structure was represented by a two-bond parameter scheme ($k = 2$ in eq. (81)),

$$A_s = E_\pi - (n_{C-C}E_{C-C} + n_{C=C}E_{C=C}) \qquad (82)$$

where the bond parameters have the following numerical values: $E_{C-C} = 0.52$ and $E_{C=C} = 2.00$. This method represents a direct transplantation of the Dewar's idea from an SCF π-MO framework to the Hückel model. Hess and Schaad [66], on the other hand, have used an eight-bond parameter scheme ($k = 8$ in eq. (81)) to represent the localized reference structure,

$$RE(\text{Hess-Schaad}) = E_\pi - \Big(n_{\text{C-C}}E_{\text{C-C}} + n_{\text{HC-C}} + E_{\text{HC-C}} +$$

$$+ n_{\text{HC-CH}}E_{\text{HC-CH}} + n_{\text{C=C}}E_{\text{C=C}} +$$

$$\tag{83}$$

$$+ n_{\text{HC=C}}E_{\text{HC=C}} + n_{\text{H}_2\text{C=C}}E_{\text{H}_2\text{C=C}} +$$

$$+ n_{\text{HC=CH}}E_{\text{HC=CH}} + n_{\text{H}_2\text{C=CH}}E_{\text{H}_2\text{C=CH}}\Big).$$

The Hess-Schaad parameters are given in Table 1.

Table 1. Comparison of perturbation results through
fourth order with empirical bond energies

Bond	Empirical Bond energy [66]	Perturbation result [95]
$C - C$	0.4358	0.4063
$HC - C$	0.4362	0.4375
$HC - CH$	0.4660	0.4688
$C = C$	2.1716	2.2500
$HC = C$	2.1083	2.1250
$H_2C = C$	2.0000[a]	2.0000
$HC = CH$	2.0699	2.0625
$H_2C = CH$	2.0000[a]	2.0000

[a]This value is arbitrarily assigned.

Schaad and Hess [95] investigated the origin of the reference structures with the two-bond and the eight-bond parameters. They treated acyclic reference structures as collections of perturbed ethylene molecules. A perturbation treatment through second order produced the two-bond parameter scheme ($E_{\text{C-C}} = 0.5;\ E_{\text{C=C}} = 2$), whereas a perturbation treatment through the fourth order produces the eight-bond parameter scheme with bond energies only slightly different from those that were determined empirically [66] (see Table 1).

These two DRE-like methods give for many compounds, very similar predictions, but there are some exceptions mainly due to the fact that a two-bond parameter scheme approximates the π electron energies of branched polyenes less accurately than an eight-bond parameter scheme. Nevertheless, both methods differentiate correctly between the aromaticity in benzenoid and non-benzenoid hydrocarbons.

There have been many attempts to produce a topological formula for the resonance energy of π-electron systems [16]. These attempts may be split into two groups. Into the first group go those attempts which aim to produce directly a topological formula for approximating RE's of conjugated molecules without worrying about the reference structure. The second group contains the efforts directed towards finding the optimum acyclic-polyene reference structure expressed in graph-theoretical terms.

Let us review the endeavours of the first group.

Green has produced a very simple topological formula for the RE of benzenoid hydrocarbons [96],

$$RE(\text{Green}) = \frac{M}{3} + \frac{m}{10} \qquad (84)$$

where m is the number of bonds contained in one benzene ring in the molecule and linking two other benzene rings. Green's formula reproduces the standard RE's of Hückel. These are valuable in the case of benzenoid hydrocarbons possessing Kekulé structures, but outside this class of conjugated structures they are useless as has already been mentioned in this section [29,66,94].

Another approximate formula for calculating RE's of benzenoid hydrocarbons has been proposed by Carter [97]. Carter's formula is given by,

$$RE(\text{Carter}) = A \cdot n_{C=C} + B \cdot \ln K - 1 \qquad (85)$$

where $n_{C=C}$ is the number of double bonds in one Kekulé structure whilst A (0.6) and B (1.5) are empirical parameters. Carter's formula could easily be modified to show the logarithmic dependence of RE on the number of Kekulé structures only,

$$RE = A \cdot \ln K. \qquad (86)$$

This was actually done by Herndon and co-workers [98] although these people at first were unfamiliar with the work of Carter. They parametrized the above equation against the Dewar-de Llano SCF π-MO resonance energies [91] and obtained for $A = 1.185$,

$$RE(SCF) = 1.185 \ln K. \tag{87}$$

Eq. (86) may be generalized to include aryl fused cyclobutadienes. In this case K should be substituted by ASC. However, a correction term for the presence of the four-membered rings should also be added to the equation. This was actually done by Wilcox [82],

$$RE(Wilcox) = A \cdot \ln (ASC) + B \cdot n_4 \tag{88}$$

where A (0.445) and B (−0.17) are least-squares parameters obtained by parametrization against the Hess and Schaad RE's [66].

McClelland [75] has also derived an approximate formula for Hückel RE's,

$$RE(McClelland) = A(2NM)^{1/2} - N + r \tag{89}$$

where $A = 0.30$, and $r = 0$ or 1, depending on whether N is even or odd.

We now turn to the second group. Attempts in this direction have been made by Hosoya *et al* [84] and by Aihara [86]. The approximate π electron energy of the reference structure in their work is given by,

$$E_\pi = A \log Z(G) \tag{90}$$

where $Z(G)$ is the Hosoya index [85]. The Hosoya index is defined as,

$$Z(G) = \sum_{n=0}^{N/2} p(G; n) \tag{91}$$

where $p(G; n)$ is the number of ways in which n bonds are chosen from G such that no two of them are connected. For acyclic structures

$$\tilde{Z}(G) = Z(G) \tag{92}$$

$$Z^*(G) = Z(G). \tag{93}$$

For cyclic structures $Z(G)$ is different from $\tilde{Z}(G)$ or $Z^*(G)$ because $Z(G)$ contains contributions from the cycles present in a molecule. If we accept that the aromaticity is a consequence of the π electronic cycles in a conjugated molecule, then the aromatic stability of conjugated systems may be predicted via the following formula,

$$RE(\text{Aihara}) = A \log \frac{Z^*(G)}{Z(G)} \tag{94}$$

or

$$RE(\text{Hosoya}) = \tilde{Z}(G) - Z(G). \tag{95}$$

Both formulae classify conjugated structures as aromatic: RE(Aihara), RE(Hosoya) > 0; non-aromatic: RE(Aihara), RE(Hosoya) = 0; and anti-aromatic: RE(Aihara), RE(Hosoya) < 0.

Finally in this section, we wish to discuss the topological resonance energy, TRE, model [16,29], which also belongs to the second group.

Topological resonance energy represents a measure of cyclic conjugation in a given polycyclic conjugated molecule. TRE is defined by [16, 29, 99, 100],

$$TRE = \sum_{i=1}^{N} (g_i x_i - g_i^{ac} x_i^{ac}) \tag{96}$$

where x_i is the i-th root of the characteristic (Hückel) polynomial, $P(G;x)$, of a conjugated molecule, whilst g_i is its occupancy number, x_i^{ac} is the i-th root of the corresponding acyclic polynomial, $P^{ac}(G;x)$, g_i^{ac} being its occupancy number. The acyclic polynomial, sometimes referred to as the reference polynomial [101] or the matching polynomial [102], is a characteristic polynomial of the Dewar-type acyclic reference structure [29]. Since in many cases,

$$g_i = g_i^{ac} \tag{97}$$

eq. (96) reduces to

$$TRE = \sum_{i=1}^{N} g_i(x_i - x_i^{ac}). \tag{98}$$

TRE is most often used in this form [29, 103]. A computer program which allows the computation of TRE's for molecules with up to 52 atoms is now available [104].

The acyclic polynomial $P^{ac}(G;x)$ is defined as,

$$P^{ac}(G;x) = \sum_{n=0}^{N} a_n^{ac}(G) x^{N-n} \tag{99}$$

and may be constructed by the selective counting of linear noncyclic elements in Sachs' method [25,30],

$$a_n^{ac}(G) = \sum_{s \in S_n^{ac}} (-1)^{c(s)}. \tag{100}$$

In the above formula s stands for the Sachs graph, set S^{ac} contains only complete graphs of degree one, i.e. K_2 graphs, whilst $c(s)$ denotes the total number of K_2 components in a given s.

The construction of the acyclic polynomial via eq. (100) becomes impractical for larger systems, even by means of the appropriate computer program, because of the enormous increase in the number of Sachs graphs. Therefore, instead of counting the Sachs graphs, we use for the (computer) generation of $P^{ac}(G; x)$, the recurrence relationship,

$$P^{ac}(G; x) = P^{ac}(G - e; \; x) - P^{ac}(G - (e); \; x). \tag{101}$$

The symbols have the following meanings: G stands for a graph of a given conjugated molecule, $G - e$ is the subgraph obtained by deletion of an edge e from G, whilst $G - (e)$ denotes the subgraph obtained by removal of an edge and incident vertices form G.

G   G − e   G − (e)

The idea behind the recurrence relationship (101) is to break up the graph G, in the minimum of steps, into linear fragments for which the polynomials are readily available [16].

In the case of hetero-conjugated systems eg. (101) must be modified in order to account for the weights of atoms and bonds. Thus, eq. (101) for the vertex- and edge-weighted (heterogeneous) graphs G_{VEW} becomes,

$$P^{ac}(G_{VEW}; x) = P^{ac}(G_{EW}; x) - h P^{ac}(G_{VEW} - v_h; x) \tag{102}$$

$$P^{ac}(G_{EW}; x) = P^{ac}(G_{EW} - e_k; x) - k^2 P^{ac}(G - (e_k); x) \tag{103}$$

where h and k are symbols for the vertex-weight and the edge-weight, $G_{VEW} - v_h$ stands for a subgraph obtained by deleting the weighted vertex v_h from G_{VEW}, G_{EW} for the edge-weighted graph, $G_{EW} - e_k$ for the subgraph obtained by removal of the weighted edge e_k from G_{EW}, whilst $G_{EW} - (e_k)$ denotes the subgraph obtained by omitting the weighted edge e_k and incident vertices from G.

$$G_{VEW} \qquad G_{EW} \qquad G_{VEW} - v_h$$

$$G_{EW} - e_k \qquad G_{EW} - (e_k)$$

Again, the idea behind eqs. (102) and (103) is to break up G_{VEW} in the minimum number of steps into linear unweighted fragments whose characteristic polynomials are rather simple.

In order to avoid the size effect the TRE is normalized. The normalization is most often carried out by dividing the TRE by number of π electrons N in the system [105],

$$TRE(PE) = \frac{TRE}{N} \tag{104}$$

where TRE(PE) stands for TRE per π electron. However, there are other possibilities open for the normalization of TRE [4,105,106]. If we normalize TRE via eq. (104) then the threshold values of TRE(PE) used for the classification of conjugated systems are as follows:

(i) molecules having TRE(PE) greater than or equal to 0.01 are considered to be of prevailing *aromatic* character;

(ii) molecules having TRE(PE) values in the interval $+0.01, -0.01$ are either ambivalent or *non-aromatic* species;

(iii) molecules having TRE(PE) less than or equal to -0.01 are viewed as *anti-aromatic*.

This classification should serve as a rough indicator for experimental chemists concerning the difficulties involved in the preparation of a given conjugated system, because a molecule labelled aromatic is usually more easily made than one labelled anti-aromatic [49–55, 87, 88, 107, 108].

An example of calculating TRE and TRE(PE) values for naphtalene is shown in Figure 6.

Figure 6. Calculation of the TRE and TRE(PE) indices of naphthalene

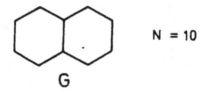

$$N = 10$$

G

(i) Hückel spectrum

$$P(G; x) = x^{10} - 11x^8 + 41x^6 - 65x^4 + 43x^2 - 9$$

$$\{2.303,\ 1.618,\ 1.303,\ 1.000,\ 0.618,$$

$$-0.618,\ -1.000,\ -1.303,\ -1.618,\ -2.303\}$$

(ii) Acyclic spectrum

$$P^{ac}(G; x) = x^{10} - 11x^8 + 41x^6 - 61x^4 + 31x^2 - 3$$

$$\{2.232,\ 1.732,\ 1.464,\ 0.865,\ 0.354,$$

$$-0.354,\ -0.865,\ -1.464,\ -1.732,\ -2.232\}$$

Figure 6. Calculation of the TRE and TRE(PE) indices of naphthalene (continued)

(iii) TRE and TRE(PE)

Energy diagram of a molecule   Energy diagram of a reference structure

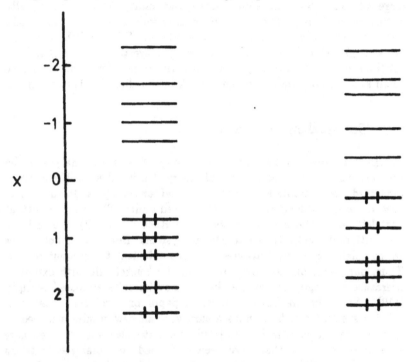

$$\sum_{i=1}^{N} g_i x_i = 13.684 \qquad\qquad \sum_{i=1}^{N} g_i x_i^{ac} = 13.294$$

$$\text{TRE} = \sum_{i=1}^{N} g_i x_i - \sum_{i=1}^{N} g_i x_i^{ac} = 0.390$$

$$\text{TRE(PE)} = \text{TRE}/10 = 0.039$$

The TRE model was independently introduced by Aihara [101,109] and it is presently used by several research groups [110–115]. However, with the increasing use of the TRE model, criticisms of it (some justified and some not) have also increased [116–118]. It appears that the majority of these critical remarks (non-physical reference structure, unacceptably large values of TRE indices for radicals, ions, radical-ions, etc.) are really criticisms of all DRE-like resonance energy indices and not only TRE, because it has been shown by several people that DRE, REPE, and TRE are strongly related aromaticity indices [4]. Nevertheless, it is clear that the TRE model must undergo a major revision in order to be applicable to all kinds of conjugated systems. Work in this direction is in progress.

6.9   Concluding Remarks

In this review we have presented a graph-theoretical analysis of the simplest variant of molecular orbital theory: Hückel theory. The analysis presented is not complete because we have left out many results in order to keep the size of the article within the agreed limits. The most important part left out is the discussion about the Hückel ($4n + 2$) rule and the application of Hückel theory to Möbius systems. However, in both cases we completed additional articles which complement the present review. Thus, our paper on the development of the Hückel rule with extensive literature coverage and some additional novel results appeared recently [119]. Similarly, we have prepared a paper on the graph theory of Möbius systems which includes a survey of all new results achieved in this area in the past few years [120]. We have also not reviewed here all reactivity indices that have been developed and analyzed in terms of graph-theoretical invariants because we believe that many of these analyses are not still definitive. In addition, we have not said anything about infinite, or near infinite, systems such as conjugated polymers, because this is a large area of research in which many people have been active [121] and in which the contribution of the Zagreb group is rather limited [122]. However, this is the area in which Hückel theory, in a classical formulation, is much less conveniently applied than its graph-theoretical formulation, and this opens up new possibilities for treating conjugate infinite systems. Therefore, as far as we can see, there is still some room left for applying graph theory to the framework of simple qualitative molecular orbital theories, because in chemistry we need, besides exact and semi-exact computational methods, simple conceptual schemes that may be used to rationalize an observed, or a computed, result in a fairly simple way [123–125].

6.10 References

1. L. Salem, *The Molecular Orbital Theory of Conjugated Systems*, Benjamin, New York 1966, p. 7.
2. E. Hückel, Z. Physik 70 (1931), 204; *ibid.* 72 (1932), 310; *ibid.* 76 (1932), 628.
3. J.N. Murrell and A.J. Harget, *Semi-empirical SCFMO Theory of Molecules*, Wiley, London 1972, Chapter 1.
4. I. Agranat, B.A. Hess, Jr. and L.J. Schaad, Pure Appl. Chem. 52 (1980), 1399; B.A. Hess, Jr. and L.J. Schaad, Pure Appl. Chem. 52 (1980), 1741; L.J. Schaad and B.A. Hess, Jr., Pure Appl. Chem. 54 (1980), 1097.
5. C.A. Coulson, B. O'Leary and R.B. Mallion, *Hückel Theory for Organic Chemists*, Academic, London 1978.
6. M. Randić and Z.B. Maksić, Chem. Rev. 72 (1972), 43.
7. B.M. Gimarc, *Molecular Structure and Bonding*, Academic, New York 1979.
8. G. Del Re, G. Berthier and J. Serre, "Electronic States of Molecules and Atom Clusters – Foundations and Prospects of Semi-empirical Methods," Lecture Notes in Chemistry, Vol. 13, Springer, Berlin 1980.
9. N.D. Epiotis, J.R. Larson and H.L. Eaton, "Unified VB Theory of Electronic Structure," Lecture Notes in Chemistry, Vol. 29, Springer, Berlin 1982.
10. Z.B. Maksić, Pure Appl. Chem. 55 (1983), 307.
11. R. Hoffmann, Nobel Lecture, Kem. Ind. (Zagreb) 32 (1983), 603.
12. A. Streitwieser, Jr., *Molecular Orbital Theory for Organic Chemists*, Wiley, New York 1961.
13. E. Heilbronner and H. Bock, *The HMO Model and Its Application. 1. Basis and Manipulation*, Wiley, London 1976.
14. M.J.S. Dewar, Chem. Britain 11 (1975), 97.
15. K. Fukui, Topics Curr. Chem. 15 (1970), 1.
16. N. Trinajstić, *Chemical Graph Theory*, CRC, Boca Raton, Florida 1983.
17. F. Harary, *Graph Theory*, Addison-Wesley, Reading, Mass. 1971, second printing.

18. L. Spialter, J. Chem. Doc. 4 (1964), 261.
19. D. König, *Theorie der endlichen und unendlichen Graphen*, Chelsea, New York 1950, reprinted, p. 170.
20. F. Sondheimer, Acc. Chem. Res. 5 (1972), 81; A.T. Balaban, M. Banciu and V. Ciorba, *Annulenes, Derivatives, and Valence Isomers*, CRC Press, Boca Raton, Florida 1987.
21. D.H. Rouvray, in *Chemical Applications of Graph Theory*, edited by A.T. Balaban, Academic, London 1976, p. 175.
22. D.M. Cvetković, M. Doob and H. Sachs, *Spectra of Graphs*, Academic, New York 1980.
23. N.S. Ham, J. Chem. Phys. 29 (1958), 1229.
24. A. Kurosh, *Higher Algebra*, Mir, Moscow 1980.
25. H. Sachs, Publ. Math. (Debrecin) 11 (1963), 119.
26. B.A. Hess, Jr., L.J. Schaad and I. Agranat, J. Am. Chem. Soc. 100 (1978), 5268.
27. I. Gutman and O.E. Polansky, Theoret. Chim. Acta 60 (1981), 203.
28. S.S. D'Amato, B.M. Gimarc and N. Trinajstić, Croat. Chem. Acta 54 (1981), 1; M. Randić, M. Barysz, J. Nowakowski, S. Nikolić and N. Trinajstić, J. Mol. Struct. (Theochem), in press.
29. I. Gutman, M. Milun and N. Trinajstić, J. Am. Chem. Soc. 99 (1977), 1692.
30. A. Graovac, I. Gutman, N. Trinajstić and T. Živković, Theoret. Chim. Acta 26 (1972), 67; N. Trinajstić, J. Math. Chem. 2 (1988), 197.
31. C.A. Coulson, Proc. Camb. Phil. Soc. 46 (1950), 202.
32. D. Cvetković, I. Gutman and N. Trinajstić, Chem. Phys. Lett. 29 (1974), 65.
33. E. Hückel, Z. Physik 83 (1933), 632; Z. Phys. Chem. B34 (1936), 335; Z. Electrochem. Angew. Phys. Chem. 43 (1937), 752; *ibid.* 43 (1937), 827.
34. R.S. Mulliken, J. Chem. Phys. 3 (1935), 375.
35. F. Bloch, Z. Physik 52 (1929), 555; *ibid.* 61 (1930), 206.
36. C.A. Coulson and A. Streitwieser, Jr., *Dictionary of π-Electron Calculations*, Freeman, San Francisco 1965.
37. W.P. Purcell and J.A. Singer, J. Chem. Eng. Data 12 (1967), 235.
38. C. Párkányi, W.C. Herndon and A.S. Shawali, J. Org. Chem. 45 (1980) 3529.
39. K. Ruedenberg, J. Chem. Phys. 22 (1954), 1878.
40. I. Gutman and N. Trinajstić, Topics Curr. Chem. 42 (1973), 49.
41. K. Ruedenberg, J. Chem. Phys. 34 (1961), 1861.

42. H.H. Günthard and H. Primas, Helv. Chim. Acta **39** (1956), 1645; N. Trinajstić, Modern Theoret. Chem. **7** (1977), 1.

43. R.B. Mallion and D.H. Rouvray, Mol. Phys. **36** (1978), 125; R.B. Mallion, Croat. Chem. Acta **56** (1983), 477.

44. C.A. Coulson, *Valence*, University Press, Oxford 1961, second edition, p. 35.

45. H.C. Longuet-Higgins, J. Chem. Phys. **18** (1950), 265.

46. E. Clar, W. Kemp and D.C. Stewart, Tetrahedron **3** (1958), 36.

47. C.R. Flynn and J. Michl, J. Am. Chem. Soc. **95** (1973), 5802.

48. I. Gutman and N. Trinajstić, Naturwiss. **60** (1973), 475.

49. W. Baker, in *Perspectives in Organic Chemistry*, edited by A. Tood, Interscience, New York 1956, p. 28.

50. *Non-Benzenoid Aromatic Compounds*, edited by D. Ginsburg, Interscience, New York 1959.

51. E. Clar, *Polycyclic Hydrocarbons*, Academic, London 1964; J.R. Dias, *Handbook of Polycyclic Hydrocarbons. Part A: Benzenoid Hydrocarbons*, Elsevier, Amsterdam 1987.

52. D. Lloyd, *Carbocyclic Non-Benzenoid Aromatic Compounds*, Elsevier, Amsterdam 1966.

53. M.P. Cava and M.J. Mitchell, *Cyclobutadiene and Related Compounds*, Academic, New York 1967; see also O.L. Chapman, C.C. Chang and N.R. Rosenquist, J. Am. Chem. Soc. **98** (1976), 261.

54. P.J. Garratt and M.V. Sargent, in *Advances in Organic Chemistry*, edited by E.C. Taylor and H. Wynberg, Vol. 6, Interscience, London 1969, p. 1.

55. *Topics in Non-Benzenoid Aromatic Chemistry*, edited by T. Nozoe, R. Breslow, K. Hafner, S. Itô and I. Murata, Vol. I, Wiley, New York 1973.

56. T. Živković, Croat. Chem. Acta **44** (1972), 351.

57. M.J.S. Dewar and H.C. Longuet-Higgins, Proc. Roy. Soc. (London) A**214** (1952), 482.

58. D. Cvetković, I. Gutman and N. Trinajstić, J. Mol. Struct. **28** (1975), 289.

59. Y.-s. Kiang, Int. J. Quantum Chem. S**14** (1980), 541; A.-c. Tang, Y.-s. Kiang, G.-s. Yan and S.-s. Tai, *Graph Theoretical Molecular Orbitals*, Science Press, Beijing 1986.

60. I. Gutman and N. Trinajstić, Croat. Chem. Acta **45** (1973), 539.

61. P. Křivka and N. Trinajstić, Coll. Czech. Chem. Comm. **50** (1985), 291.

62. D. Cvetković, I. Gutman and N. Trinajstić, Chem. Phys. Lett. 16 (1972), 614.
63. C.F. Wilcox, Jr., Tetrahedron Lett. (1968), 795.
64. W.C. Herndon, Tetrahedron 29 (1973), 3.
65. C.A. Coulson and G.S. Rushbrooke, Proc. Cambridge Phil. Soc. 36 (1940), 193.
66. B.A. Hess, Jr. and L.J. Schaad, J. Am. Chem. Soc. 93 (1971), 305.
67. M.J.S. Dewar, *The Molecular Orbital Theory of Organic Chemistry*, McGraw-Hill, New York 1969.
68. L.J. Schaad and B.A. Hess, Jr., J. Am. Chem. Soc. 94 (1972), 3068.
69. C.A. Coulson, Proc. Cambridge Phil. Soc. 36 (1940), 201.
70. R.D. Brown, Trans. Faraday Soc. 46 (1950), 1016.
71. C.A. Coulson, J. Chem. Soc. (1954), 3111.
72. G.G. Hall, Proc. Roy. Soc. (London) A229 (1955), 251.
73. I. Gutman and N. Trinajstić, Chem. Phys. Lett. 17 (1972), 535.
74. I. Gutman, Theoret. Chim. Acta 35 (1974), 355.
75. B.J. McClelland, J. Chem. Phys. 54 (1971), 640.
76. I. Gutman, M. Milun and N. Trinajstić, J. Chem. Phys. 59 (1973), 2772.
77. I. Gutman, Chem. Phys. Lett. 24 (1974), 283.
78. I. Gutman, J. Chem. Phys. 66 (1977), 1652.
79. D. Bonchev and N. Trinajstić, J. Chem. Phys. 67 (1977), 4517.
80. G.G. Hall, Int. J. Math. Educ. Sci. Technol. 4 (1973), 223.
81. I. Gutman, N. Trinajstić and C.F. Wilcox, Jr., Tetrahedron 31 (1975), 143.
82. C.F. Wilcox, Jr., Croat. Chem. Acta 47 (1975), 87.
83. I. Gutman and N. Trinajstić, Z. Naturforsch. 29a (1974), 1238.
84. H. Hosoya, K. Hosoi and I. Gutman, Theoret. Chim. Acta 38 (1975), 37.
85. H. Hosoya, Bull. Chem. Soc. Japan 44 (1971), 2332.
86. J.-i. Aihara, J. Org. Chem. 41 (1976), 2488.
87. I. Agranat, in *Aromatic Compounds, MTP International Review of Science*, edited by H. Zollinger, Butterworths, London 1973, p. 139.
88. D. Lewis and D. Peters, *Facts and Theories of Aromaticity*, Macmillan, London 1975.
89. N. Trinajstić, Croat. Chem. Acta 37 (1965), 307.
90. N.C. Baird, J. Chem. Educ. 48 (1971), 509.
91. M.J.S. Dewar and C. de Llano, J. Am. Chem. Soc. 91 (1969), 789; M.J.S. Dewar and N. Trinajstić, J. Chem. Soc. A (1969), 1754; M.J.S. Dewar and N. Trinajstić, Tetrahedron Lett. (1969), 2129;

M.J.S. Dewar, A.J. Harget and N. Trinajstić, J. Am. Chem. Soc. 91 (1969), 6321; M.J.S. Dewar and N. Trinajstić, J. Am. Chem. Soc. 92 (1970), 1453; M.J.S. Dewar and N. Trinajstić, Croat. Chem. Acta 42 (1970), 1; M.J.S. Dewar and N. Trinajstić, Tetrahedron 26 (1970), 4269; M.J.S. Dewar and N. Trinajstić, Theoret. Chim. Acta 17 (1970), 235; M.J.S. Dewar, M.C. Kohn and N. Trinajstić, J. Am. Chem. Soc. 93 (1971), 3437; N. Tranajstić, J. Mol. Struct. 8 (1971), 236; N. Trinajstić, Record Chem. Progress 32 (1971), 199.

92. B.A. Hess, Jr. and L.J. Schaad, J. Org. Chem. 37 (1972), 419; I. Gutman, M. Milun and N. Trinajstić, Chem. Phys. Lett. 23 (1973), 284.

93. B.A. Hess, Jr. and L.J. Schaad, J. Am. Chem. Soc. 93 (1971), 2413; *ibid.* J. Org. Chem. 36 (1971), 3418.

94. M. Milun, Ž. Sobotka and N. Trinajstić, J. Org. Chem. 37 (1972), 139.

95. L.J. Schaad and B.A. Hess, Jr., Israel J. Chem. 20 (1980), 281.

96. A.L. Green, J. Chem. Soc. (1956), 1886.

97. P.C. Carter, Trans. Faraday Soc. 45 (1949), 597.

98. R. Swinborne-Sheldrake, W.C. Herndon and I. Gutman, Tetrahedron Lett. (1975), 755.

99. P. Ilić, B. Sinković and N. Trinajstić, Israel J. Chem. 20 (1980), 258.

100. P. Ilić and N. Trinajstić, J. Org. Chem. 45 (1980), 1738.

101. J.-i. Aihara, J. Am. Chem. Soc. 98 (1976), 2750.

102. E.J. Farrell, J. Comb. Theory B27 (1979), 75.

103. N. Trinajstić, Int. J. Quantum Chem. S11 (1977), 469; M. Milun and N. Trinajstić, Croat. Chem. Acta 49 (1977), 107; I. Gutman, M. Milun and N. Trinajstić, Croat. Chem. Acta 49 (1977), 441; A. Sabljić and N. Trinajstić, J. Mol. Struct. 49 (1978), 415; Croat. Chem. Acta 51 (1978), 249; P. Ilić and N. Trinajstić, Pure Appl. Chem. 52 (1980), 1495; Croat. Chem. Acta 53 (1980), 591; P. Ilić, A. Jurić and N. Trinajstić, Croat. Chem. Acta 53 (1980), 587; A. Sabljić and N. Trinajstić, J. Org. Chem. 46 (1981), 3457; B.M. Gimarc and N. Trinajstić, Inorg. Chem. 21 (1982), 21; P. Ilić, B. Mohar, J.V. Knop, A. Jurić and N. Trinajstić, J. Heterocycl. Chem. 19 (1982), 625; A.H. Stollenwerk, B. Kanellakopulos, H. Vogler, A. Jurić and N. Trinajstić, J. Mol. Struct. 102 (1983), 377; A. Jurić, A. Sabljić and N. Trinajstić, J. Heterocycl. Chem. 21 (1984), 273; P. Ilić, B. Sinković and N. Trinajstić, Theochem. 136 (1986), 155; A. Jurić, N. Trinajstić and G. Jashari, Croat. Chem. Acta 59 (1986), 617; S. Nikolić, A. Jurić and N. Trinajstić,

Heterocycles 26 (1987), 2025; Ž. Saničanin, A. Jurić, I. Tabaković and N. Trinajstić, J. Org. Chem. 52 (1987), 4053.

104. B. Mohar and N. Trinajstić, J. Comput. Chem. 3 (1981), 28.
105. P. Ilić, B. Džonova-Jerman-Blažič, B. Mohar and N. Trinajstić, Croat. Chem. Acta 52 (1979), 35.
106. P. Ilić and N. Trinajstić, Croat. Chem. Acta 56 (1983), 203.
107. G.M. Badger, *Aromatic Character and Aromaticity*, University Press, Cambridge 1969.
108. P.J. Garratt, *Aromaticity*, McGraw-Hill, New York 1971.
109. J.-i. Aihara, J. Am. Chem. Soc. 99 (1977), 2048; Bull. Chem. Soc. Japan 51 (1978), 1788; *ibid.* 51 (1978), 3540.
110. U. Norrinder, D. Tanner, B. Thulin and O. Wennerström, Acta Chem. Scand. B35 (1981), 403.
111. U. Norrinder, D. Tanner and O. Wennerström, Croat. Chem. Acta 56 (1983), 269.
112. K. Jug, J. Org. Chem. 48 (1983), 1344.
113. S. El-Basil, Indian J. Chem. 20B (1981), 586.
114. I.M. Mladenov and J.B. Vassileva-Popova, Int. J. Quantum Chem., in press.
115. J.P. Gastmans, D.F. Gastmans and T.M. Lopez, Bull. Soc. Chim. France (1983), II, 274.
116. I. Gutman, Chem. Phys. Lett. 66 (1979), 595; Theoret. Chim. Acta 56 (1980), 89; J.C.S. Faraday Trans. II (1983), 337.
117. I. Gutman and B. Mohar, Chem. Phys. Lett. 69 (1980), 375.
118. E. Heilbronner, Chem. Phys. Lett. 85 (1982), 377.
119. D.J. Klein and N. Trinajstić, J. Am. Chem. Soc. 106 (1984), 8050.
120. N. Trinajstić, to be published.
121. A. Graovac, O.E. Polansky and N. Tyutyulkov, Croat. Chem. Acta 56 (1983), 325; see also many papers by Professor Tyutyulkov on conjugated linear structures that appeared in the last decade.
122. A. Graovac, I. Gutman, M. Randić and N. Trinajstić, Colloid Polymer Sci. 255 (1977), 1692; A. Graovac, M. Randić and N. Trinajstić, Croat. Chem. Acta 53 (1980), 587.
123. O.E. Polansky and M. Zander, J. Mol. Struct. 84 (1982), 361; O.E. Polansky, M. Zander and I. Motoc, Z. Naturforsch. 38a (1983), 196; W. Fabian, I. Motoc and O.E. Polansky, Z. Naturforsch. 38a (1983), 916; N. Tyutyulkov, O.E. Polansky, P. Schuster, S. Karabunarliev, and C.I. Ivanov, Theoret. Chim. Acta 67 (1985), 211; N. Tyutyulkov, P. Schuster and O.E. Polansky, Theoret. Chim. Acta 63 (1983), 291; I. Gutman, A. Graovac and O.E. Polansky, Chem. Phys.

Lett. **116** (1985), 206; O.E. Polansky, in *Mathematics and Computational Concepts in Chemistry*, edited by N. Trinajstić, Horwood, Chichester 1986, p. 262; O.E. Polansky, in *MATH/CHEM/COMP 1988*, edited by A. Graovac, Elsevier, Amsterdam, in press.
124. A.T. Balaban, J. Mol. Struct. (Theochem) **120** (1985), 117.
125. N. Trinajstić, CRC Rev. Theoret. Chem. Biophys., in press.

INDEX